AF761801

Neuroscience of the Nonconscious Mind

Neuroscience of the Nonconscious Mind

Rajendra D. Badgaiyan
Chief of Psychiatry, South Texas Veteran Health Care System,
San Antonio, TX, United States

Professor of Psychiatry, Joe & Teresa Lozano Long
School of Medicine, University of Texas Health Science Center,
San Antonio, TX, United States

ELSEVIER

ACADEMIC PRESS
An imprint of Elsevier

Academic Press is an imprint of Elsevier
125 London Wall, London EC2Y 5AS, United Kingdom
525 B Street, Suite 1650, San Diego, CA 92101, United States
50 Hampshire Street, 5th Floor, Cambridge, MA 02139, United States
The Boulevard, Langford Lane, Kidlington, Oxford OX5 1GB, United Kingdom

British Library Cataloguing-in-Publication Data
A catalogue record for this book is available from the British Library

Library of Congress Cataloging-in-Publication Data
A catalog record for this book is available from the Library of Congress

ISBN: 978-0-12-816115-9

For Information on all Academic Press publications
visit our website at https://www.elsevier.com/books-and-journals

Publisher: Nikki Levy
Acquisition Editor: Joslyn Chaiprasert-Paguio
Editorial Project Manager: Pat Gonzalez
Production Project Manager: Sujatha Thirugnana Sambandam
Cover Designer: Matthew Limbert

Typeset by MPS Limited, Chennai, India

Contents

Preface

Scientific study of consciousness began a few years ago. Although many laboratories are trying to understand its biological basis, nobody knows for sure what consciousness means in scientific terms. So, how can we study a phenomenon that cannot even be defined and why it is so difficult to define something that almost everybody—scientists and nonscientists alike—knows about? That is precisely the problem. We all know what consciousness is, but everybody has a different definition for it and science requires a definition that is logical and acceptable to the scientific community. That has not happened so far.

Can we really talk about the nonconscious mind yes, if we do not have an acceptable definition of consciousness? The answer is, we can, but we have to first agree on what the nonconscious mind is. In this book, I have used the "common sense" definition of both terms: "nonconscious" and "mind." We use the term nonconscious to describe all actions that we perform without being consciously aware of them. For example, moving legs while walking and moving fingers while typing. We are aware of walking and typing but not of taking each step forward or moving each finger to tap a key—unless you are walking or typing for the very first time! Several other examples will be discussed in the following chapters. For actions like these, we will *not* use the term unconscious even though many philosophers, cognitive scientists, textbooks, and almost all of the older literatures have used the terms nonconscious and unconscious interchangeably. I make a distinction between these terms.

I also consider a task nonconscious if it is performed without conscious awareness even if a person is otherwise conscious. While we are not usually consciously aware of movements we make while walking and typing, we perform those actions under full consciousness. We are however, able to voluntarily control some of those functions (but not all). In these cases the function is no longer nonconscious. For example, while typing we can consciously move fingers to respective keys and while walking move legs at will to take each step but if we do so these functions are no longer nonconscious ones. A function can therefore be both conscious and nonconscious depending on how it is performed. Unconscious, on the other hand, refers to a state in which a person is incapable of performing a task at will. For example, an individual under general anesthesia or in a coma is incapable of performing a task at will. We make this distinction because the neural networks involved in processing are different under nonconscious and unconscious conditions. Since many researchers have not made a distinction between the terms nonconscious, unconscious, and sometimes the subconscious, we have tried to use their original terminology in this text to avoid distorting their meanings. However, we only discuss those concepts of the "subconscious" and "unconscious" mind that are similar to the above definition of the nonconscious mind.

Mind is another term we need to define. In the context of this book this term refers to higher mental functions like memory, attention, language, and executive processing. We have also included emotion, extrasensory perception, decision making, dreams, and hypnosis under the broad definition of mind. While we I will talk about consciousness but will not discuss the soul. I do not know what the soul is neurobiologically, and cannot even guess whether it exists in the neuroscience domain.

Now coming back to nonconscious mental tasks, I have been asked numerous times in formal scientific meetings and in informal gatherings how can we perform a mental task without being aware of it. Are not we aware of what we do? The answer is no. We perform mental functions more often without conscious awareness than with awareness. I say this with a strong conviction because all conscious actions include some form of nonconscious mental processing. For example, conversation requires conscious action to decide the subject matter, but modulation of voice, selection of the words, grammar, and syntax that goes with it are processed nonconsciously. Nonconscious processes put the conversation in the right context, recall past experiences, retrieve memories of linguistic learning, and select appropriate words to use. Thus, a great deal of nonconscious processing is needed to execute seemingly simple conscious tasks.

For another example, try to remember how many times you drove a car on a familiar route while enjoying music, talking on the phone, or simply thinking about work or home? How many times were you unable to recall something on or around the road? Probably many times. So, what is happening here? We drive consciously but still do not remember everything such as when we applied the breaks or moved the steering wheel. This means the brain must have processed everything that was on the road and also recalled driving lessons and traffic rules. Since driving involves complex information processing both at spatial and temporal levels, we process information skillfully at a very high level while driving, but do not remember most driving events. How is this possible? There are many possible reasons: one, stimuli were processed without awareness; two, they were processed with full awareness but quickly vanished from memory; three, perception never entered memory; and four, memories of stimuli were not retrievable. These possibilities sometimes make it difficult to understand the neural processing of nonconscious mind. Any of the above possibilities could account for processing of information without awareness.

In our daily lives, we perform many functions, without consciously aware of them. While changing clothes, while typing and walking, as mentioned above, etc. Similarly, we walk with full conscious awareness but we do not remember every step and every pot hole or rock avoided. So do we need to be consciously aware of a stimulus before processing and responding to it, or does the brain do it for us? After all, the brain controls vital processes like circulation and respiration without our conscious awareness. It also makes changes such as increasing heart rate during exercise, for example. Of course, we can control some of these processes like respiration but only within limits. We cannot voluntarily stop breathing indefinitely! In this context, it is also reasonable to ask whether the brain controls cognition, behavior, emotion and all of our actions nonconsciously and provides us limited ability to modify them at will? I think there is no reason to believe why it might not be the case. Most of our actions are probably executed nonconsciously without our knowledge and without conscious awareness. It is therefore important to understand nonconscious mind.

In the following chapters I have discussed how the nonconscious mind makes us smarter and more resilient and how it provides survival benefits. It does not mean the conscious mind is not important. It is. There is a place for both. In the book I have discussed how the conscious and nonconscious mind interact to make us the resilient and intelligent species we are.

Chapter 1

Historical perspective

It is indeed surprising that we do not have acceptable definitions of the conscious and nonconscious mind even though these concepts were developed thousands of years ago in ancient civilization. Ancient Indian literature, the *Vedas*, which originated in oral form around 4000 BCE (but scripted around 1500 BCE) have detailed descriptions of the conscious and nonconscious mental functions.[1] These concepts, however, remained unappreciated for centuries possibly because scriptures were in Sanskrit, an old and dead language. The Vedas have now been translated into many languages. The basic concepts of consciousness described in Vedas are as relevant today as they were centuries ago.

The following is a brief snapshot of those concepts. I have spelled Sanskrit words to sound as close to the original phonation as possible. Therefore, spellings in this book may differ from those used by other authors. Additionally, because ancient Indian scriptures evolved over centuries, there are several versions and schools of thought with slightly different concepts. For the sake of simplicity I have discussed only those concepts that are accepted by most scholars. *Vedas* are a collection of many scriptures and the word literally means "knowledge" in Sanskrit. Therefore *Vedas* are considered books of knowledge.

The *Vedas* describe two forms of consciousness: universal consciousness called *brahman* and personal consciousness called *atman*.[1–3] These two forms are practically indistinguishable because *atman* is a part of *brahman* (Fig. 1.1). There is only one *brahman* but each individual has *atman*. Therefore there is no difference in *atman* of two people. It imparts basic consciousness, which is same in all members of the human species. What distinguishes people is their inner self or *antahkaran*, which is a form of *atman*. This inner self has four components: *ahankar* (ego), *buddhi* (intellect), *manas* (senses), and *chitta* (mind). Further, *chitta* or mind consists of five entities: *jagrat chitta* (wakeful consciousness), *sanskar chitta* (subconscious mind), *vasana chitta* (subsubconscious mind), *karan chitta* (superconscious mind), and *anukaran chitta* (subsuperconscious mind). Thus Vedic literature describes two forms of nonconscious mind: *sanskar chitta* and *vasana chitta*. *Sanskar* is the upper layer of subconscious and is a stage just beneath conscious layer. It is supposed to be a repository of past experience. *Vasana*

Neuroscience of the Nonconscious Mind. DOI: https://doi.org/10.1016/B978-0-12-816115-9.00001-2

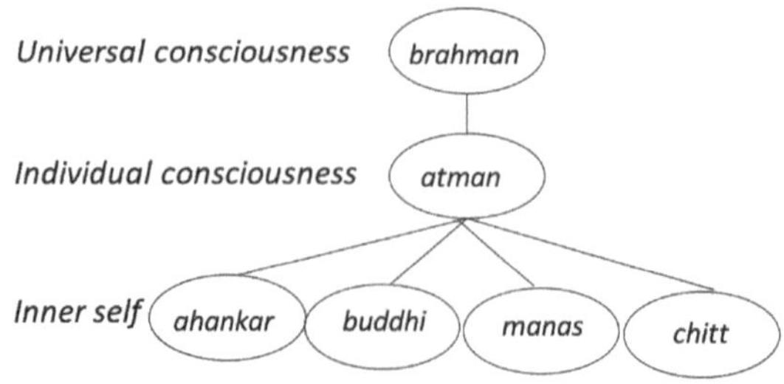

FIGURE 1.1 Schematic representation of a simplified *vedic* concept of consciousness.

chitta operates at a deeper level and it provides a framework for the mind to work. It defines a person's personality by guiding his or her thoughts and actions.

These concepts are not much different from those of the modern western philosophy developed in the 19th century after publication of the book *Philosophy of the Unconscious*.[4] The original German version was published by philosopher Karl Robert Eduard von Hartmann (Fig. 13.1) in 1869 and an English translation appeared in 1884. Hartmann bundled vedic concepts with contemporary German philosophy and proposed three forms of nonconscious/unconscious mind: absolute unconscious, which is a substance of the universe and is the source of all other unconscious; physiological unconscious, which is at work in the origin, development, and evolution of living beings; and psychological unconscious, which lies at the source of our conscious mental life. This concept was further developed by Austrian neurologist Sigmund Freud (Fig. 1.6). In an article titled "The Unconscious,"[5] published in 1915, Freud suggested that the unconscious mind primarily works as a repository of information and has no processing function. He developed psychoanalytical techniques to bring nonconscious information to conscious awareness and used them to treat psychological disturbances. It is now clear that the nonconscious mind is not merely a repository; it performs high-level cognitive processing while keeping it out of our conscious awareness. Freud also suggested that the unconscious mind modifies our actions based on repressed desires, drives, and motivations.

Since a discussion of philosophical concepts is out of the scope of this book, these concepts will not be discussed. Here, we focus on scientific evidence and concepts.

Evidence of the existence of nonconscious mind was experimentally demonstrated for the first time in 1884 by an American logistician named Charles S. Peirce (Fig. 1.2) at Johns Hopkins University. Peirce was admired as one of the greatest logisticians, philosophers, and mathematicians of the time, but because of his antipathy to the then president of Harvard University Charles Elliot, he could not get a faculty position there. He worked as a lecturer at Johns Hopkins but was dismissed because of an extramarital affair. Peirce died in poverty and lived on donated money raised

FIGURE 1.2 Charles Peirce (1839–1914). Conducted experiments on the nonconscious mind. *Reprinted from Wikimedia Commons.*

FIGURE 1.3 Boris Sidis (1867–1923). His experiments provide scientific evidence of the existence of nonconscious mind. *Reproduced from Wikimedia Commons.*

by his friend William James (Fig. 7.1), a well-known psychologist at Harvard. After death, Peirce left 1650 unpublished manuscripts in over 100,000 pages.

Peirce conducted a landmark experiment with Joseph Jastrow (1863–1944). In this experiment he asked volunteers to make an estimate of the weight placed on a pane of a balance and also to estimate the degree of confidence in their answer. He found that despite having very low confidence, estimates were close to the actual weight.[6] This observation made him believe that volunteers "knew" something that they were not consciously aware of. Peirce conducted another experiment in which volunteers estimated the luminosity of a lamp. He found similar results.

With these experiments, Peirce and Jastrow introduced the nonconscious mind to experimental psychology, but the first major contribution in this area was made by a psychiatrist Boris Sidis (Fig. 1.3). He described most of his work in the book *The Psychology of Suggestion*, published in 1898.[6,7] Sidis was born in Ukraine and immigrated to the United States in 1887 to avoid political persecution. After receiving PhD and MD degrees from Harvard University, he joined the faculty of its psychology department. Sidis named his son William James in honor of his friend and the famous psychologist who had the same name. Junior Sidis was consciously raised to be a prodigy. As a result, he was said to be proficient in 40 languages and earned a BA degree from Harvard at age 16. However, those were the only extraordinary achievements of his life. He died at the age of 46 without making a lasting intellectual contribution.

Sidis published many groundbreaking studies on nonconscious perception. He first experimented on hypnotized volunteers and then on himself and finally on nonhypnotized people. He observed that the threshold for visual, auditory, and haptic perception is significantly reduced under hypnosis. They therefore can perceive stimuli that cannot normally be perceived. This observation led him to extend the study on himself. Since Sidis had an amblyopic right eye he was particularly interested in visual perception. With the right eye he could not differentiate whether a printed character was a letter or a number. While experimenting with this eye he tried to guess whether a card had a number or letter printed on it. He found that he could guess correctly significantly above the chance level. Believing that he was perceiving sensory stimuli nonconsciously, he repeated those experiments on healthy volunteers. They were shown cards that had either a number or a letter written on them. Cards were placed at a distance from where volunteers could not read and would see the writing as "dim blurred spots." They were then asked to guess if the writing was a letter or a number and also to guess which letter or number was printed on the card. To his surprise, volunteers correctly guessed the category (number or letter) in 70% of trials and correctly identified the number or letter in 34% of trials. It was much higher than expected and could not be explained by chance alone. This observation suggested that even though volunteers could not read the print, their brains somehow read it without making them consciously aware of it.

Sidis' next experiment was equally fascinating. In this experiment he asked volunteers to look at a complicated drawing printed on a card for 10 seconds and then to reproduce it. He also asked them to reproduce one of eight numbers printed on top of the card. The card had another number in the margin. Even though volunteers denied seeing them, 32% of volunteers chose those numbers instead of the one on the top. Based on these observations Sidis developed a psychophysiological theory that suggests that the brain has a second channel for receiving information. From this channel information is selectively passed on to the "upper consciousness." This concept is remarkably similar to the one we developed recently using neuroimaging data. Our concepts are described in Chapter 2, Nonconscious memory.

Sidis' experiments suggested that we perceive more than what we know. It led investigators to conduct experiments to determine whether stimuli that cannot be perceived affect our cognition and behavior. Most of these experiments used subliminal stimuli, which are visual stimuli presented briefly for a few milliseconds. Because of the short duration of presentation, stimuli are not consciously perceived. Before in-depth research on cognitive and behavioral influences of these stimuli could be established, subliminal stimuli became a subject of controversy because of allegations of its use by commercial and government entities to alter people's opinions and behaviors. The most infamous controversy was created by a Detroit-born market researcher James McDonald Vicary (Fig. 1.4). At age 13, Vicary rose to fame as the youngest snake charmer for his ability to comfortably handle

FIGURE 1.4 Top: James McDonald Vicary (1915–77) as a young snake charmer. He later falsely claimed that subliminal stimuli affect our behavior. Snapshot of a clip originally published in the *Detroit News*. Bottom: An example of a subliminal camouflaged advertisement. The images at the top right show a magnified view of the camouflaged word "sex" in an advertisement. *Image on the top reproduced with permission from Global ImageWork and the one on the bottom reproduced from Open Library.*

snakes. But his real fame began with an "experiment" he conducted in 1957. He claimed that in a Fort Lee, New Jersey movie theater, he repeatedly showed a message "Hungry? Eat Popcorn" and "Drink Coca Cola" for 1/3000th of a second. After this subliminal message, the sale of popcorn increased 57.5% and that of Coca Cola increased 18.1%.[8] It generated a lot of public interest even though his data were challenged by many investigators. When Vicary was forced to repeat the experiment, he was not able to replicate the results and manager of the Fort Lee theater denied any knowledge of such an experiment ever conducted in the theater.

Despite doubts about veracity of Vicary's claims, radio and television stations began airing subliminal commercials. This led to introduction of at least two bills in the US Congress to ban subliminal stimuli in 1958 and 1959 (both bills died before voting). Later, in a 1962 television interview, Vicary admitted that the Fort Lee experiment was a gimmick. Despite Vicary's confession, fascination and commercialization of these stimuli continued unabated. Subliminal stimulus was exploited for commercial purposes by a journalism professor Wilson Bryan Key (1925–2008) who advanced the subliminal story by writing several books and suggesting different ways to use it in advertising. In one of his popular books *Subliminal Seduction*,[9] he claimed that an advertisement can be made attractive and enticing if it includes the camouflaged word "sex." According to him, the word needed to be camouflaged to prevent conscious perception (Fig. 1.4). He argued that because sex is a stimulating concept, it would make the advertisement attractive and promote sale of the product if the word was perceived nonconsciously. Publication of his book on subliminal advertising prompted the Federal Communications Commission of the United States in 1974 to ban subliminal stimuli in advertising. However, it did not stop print media from continuing to use hidden messages such as the one shown here in an advertisement for Coca Cola in Australia (Fig. 1.4). It also led to commercialization of a number of self-help subliminal tapes that claimed to help people with obesity, smoking, memory loss, sexual problems, etc. When these tapes were scientifically evaluated, none of the claims could be verified. In a study by Anthony Greenwald of Washington University volunteers listened to tapes that claimed either improved memory or self-esteem. After a month of listening there was no improvement in any of the claimed attributes.[10]

In the early 1980s American radio and television stations engaged in spirited debate on the effects of subliminal messages. One particular type of message that received the most attention was the backmasking, a term used to describe a process in which a message is recorded backward and inserted into a music album. While playing the album the message was not comprehensible but when played backward it could be clearly heard. It was widely believed at the time that increasing incidence of violence, drug use, and sexual promiscuity in society was because of hidden back-masked messages in rock music. A well-known pastor from California, Gary Greenwald, was in particular known to travel all over the United States and Canada to propagate this idea. He created intense public outcry forcing California legislature to pass a bill banning backmasking without public notice. The state of Arkansas required printing of a warning on any record using backmasking: "This record contains backward masking, which may be perceptible at a subliminal level when the record is played forward."

The belief in the negative effects of backmasking gained additional momentum after two young men from Reno Nevada, named Raymond Belknap (18 years) and James Vance (20 years), shot themselves at a local playground on December 23, 1985 after using drugs and listening to the album "Stained Glass" recorded by the British heavy metal band Judas Priest. Belknap died instantly and Vance 3 years later. Their families sued Judas Priest and CBS records, holding them responsible for their deaths. The trial went on for a month, and it was claimed that the album had the subliminal messages: "let's be dead" and "do it." The defendants denied the accusation and noted that if they were to put a message on their albums they would have preferred a message asking listeners to buy copies of the album, not to kill themselves. Interestingly, it was found that the album did include hidden messages. When it was played backward, phrases like, "hey ma, my chair is broken," "give me a peppermint," and "help me keep a job" were heard. The judge dismissed the case by saying that "the research data presented does not establish that subliminal stimuli, even if perceived, may precipitate conduct of this magnitude...." After this case, many investigators examined the effect of buried subliminal messages. None of them ever found any evidence of a significant effect.[11]

Because of the adverse media attention and fraudulent claims, nonconscious perception was considered taboo in scientific field. It was further discounted because of methodological flaws in initial experiments and because many investigators were unable to replicate results of their own experiments. One of such studies was reported by Lynn E. Baker of the University of Wisconsin.[12] In this study, she claimed that pupillary light reflex can be conditioned by the presentation of tone that cannot be perceived consciously as unconditioned stimulus. Baker also claimed that conditioning occurs only in three trials and it is difficult to extinguish. Since she made bold claims many investigators tried to replicate her results but were unable to establish conditioned pupillary light reflex with subliminal stimulus. Another issue with experiments on nonconscious perception was the validity of experimental protocols. Investigators who found an effect of these stimuli did not use objective measures to determine subliminal perception and relied on a subject's subjective report. Moreover, these experiments did not take into account false alarm, variation in perceptibility, effect of expectation, and distinction between perception and verbalization.[13]

Because of these issues studies on nonconscious stimuli in general, and subliminal stimuli in particular, were not considered reliable and most investigators avoided studying these stimuli. However, they continued to study a related phenomenon called dichotic listening. In these studies, volunteers were presented different auditory stimuli in each ear and asked to attend only to the one heard in a particular ear. For example, if they heard the word "table" in the right ear and "chair" in the left ear they would be asked to attend to what they heard in the right ear and ignore words presented to the left. In these experiments, even though ignored stimuli were not consciously attended to, they influenced volunteers' understanding of attended stimuli.[14] Thus if ignored stimulus is a word "nurse," it is easier for volunteers to attend to word "doctor" presented in the attended ear. This effect has been observed when two words (in this case, nurse and doctor) are semantically related. A similar effect was observed in a visual version of this experiment (dichoptic vision). In these experiments, pictures of different objects were shown in the two eyes and volunteers were asked to focus on the picture presented to a specific eye and ignore that presented to the other eye. These studies suggested that ignored stimuli are semantically processed even though they are not consciously attended to.[15] This effect was observed even when stimuli presented to the unattended eye interfered with the attended stimuli. In this condition volunteers had no incentive to attend to the stimulus presented to the unattended eye.[16] Dichotic and dichoptic experiments demonstrated that the brain processes even those stimuli that we do not pay attention to and do not even remember. These findings indicate that perception, cognition, and conscious awareness are separate and dissociable processes.

This dissociation was further demonstrated by a British psychologist named Anthony Marcel who demonstrated that subliminally presented words are processed cognitively even though they are not consciously perceived. He demonstrated the dissociation between cognition and perception by showing that subliminal words facilitate identification of a semantically associated word. Marcel showed this effect, called semantic priming, in a series of meticulously planned experiments using objective criteria. Most significantly, his experiments were replicated in many laboratories. Marcel's work provided legitimacy to nonconscious perception, particularly because he used an established psychological tool to study subliminal stimulus. His experiments were described in two influential papers he published in 1983[17,18] only 3 years after obtaining a PhD degree from the University of Reading (he is currently a professor emeritus at the University of Hertfordshire and Cambridge University). In the 1980s, the concept of semantic priming was well established and many investigators had shown that processing of a word is quicker if it is primed by a semantically related word. Thus if a subject is shown the word "nurse" (prime word), subsequent presentation of the word "doctor" results in quicker response because the two words are semantically related. In his experiments, Marcel demonstrated a similar effect when subjects were primed with subliminally presented words. In an experiment he presented two probe words and asked volunteers to guess which of the two probes was semantically similar to the subliminally presented word. He found that the response was quicker when the subliminal word was semantically related to the probe. Based on these findings, he proposed that "perceptual processing itself is nonconscious and automatically proceeds to all levels of analysis."

Marcel's claim that our conscious behavior is influenced by nonconsciously perceived stimuli sparked both interest and controversy. Most of the controversy originated from psychoanalysts who objected to the use of terms nonconscious and unconscious in this context because it was not consistent with their philosophical definitions. The controversy eventually led to evolution of the idea that these terms can be used by both psychoanalysts and cognitive scientists—perhaps in a slightly different context. The scientific community at first did not approve that the nonconscious and unconscious mind can coexist in psychoanalysis and cognitive science. Only a handful of scientists supported this idea; they believed in so-called new look theory proposed in 1947 by Bruner and Postman.[19] Since the new look theory did not get much attention, a Hungarian-American psychologist named Matthew Hugh Erdelyi (professor of psychology at the City University of New York) attempted to revive it in 1974 by establishing a theoretical connection between psychoanalytical and scientific concepts of unconscious cognition.[20] He argued that information processing theory (multistore information processing models of attention and memory, which was popular at the time), allows some degree of perception without awareness by recognizing existence of preattentive processing and short-term memory. Despite making a strong theoretical argument, Erdelyi was also not very effective in convincing the scientific community of nonconscious cognition.

Primarily because of skepticism, data acquired using subliminal stimuli were criticized both at theoretical and methodological levels. Theoretical criticism was primarily based on the belief that a stimulus has to be consciously perceived before it can be semantically processed for its meaning.[21] The idea that a stimulus can be semantically processed without conscious perception was not consistent with models of cognition prevalent at the time.[22] Supporters of those models suggested that methodological issues might have led investigators to arrive at erroneous conclusions in experiments involving subliminal stimuli.

Since Marcel used a well-established and tested method, criticism of his experiments was focused on the claim of "perception without awareness." Some of the questions raised were legitimate. For example, how can one be sure that the stimuli presented in his experiments were subliminal, and how can we rule out the possibility that stimuli were perceived consciously but were not registered in memory? These are valid questions because there is no objective measure of subliminal exposure, and the time for which a stimulus must be exposed to make it subliminal depends on a number of variables including type of stimulus, state of the eyes (dark or light adapted), type of test (semantic or perceptual), acuity of vision, and many other factors that might include level of attention paid and psychological state of mind. In Marcel's experiments the duration of subliminal exposure varied between 20 and 110 ms. He estimated a subliminal cutoff time of each volunteer individually by presenting the prime stimulus for different time periods and determining the time at which the volunteer's response was at chance level (50% correct response if there were two choices). The maximum exposure time that produced chance level performance was considered the subliminal cutoff time for that volunteer. Absence of an objective measure to verify subliminal exposure is a reasonable criticism of Marcel's experiments because if individual volunteers in an experiment perform at a chance level between 20 and 110 ms, then why wasn't exposure time fixed at 20 ms for all volunteers? Instead. Marcel used different exposure times for different volunteers depending on their subliminal cutoff time. Thus the exposure time used was at the border of subliminal and supraliminal perception. This strategy could have made at least some stimuli supraliminal because the subliminal threshold could vary from stimulus to stimulus as mentioned above.

These criticisms were validated by the fact that subliminal stimuli do not elicit any cognitive effect if exposure time is too short. Absence of an acceptable threshold for subliminal exposure led two Canadian psychologists, Jim Cheesman and Philip Merikle, to differentiate between objective and subjective thresholds.[23] Objective threshold is the exposure time that yields chance level of performance used by Marcel. They defined the subjective threshold as the time at which a person does not consciously recognize a stimulus. Evidence of semantic priming is observed only if exposure is above the objective threshold. This could suggest that "subliminal stimuli" are consciously perceived, but because of relatively short exposure time, the stimuli are not registered in memory. Marcel's experiments were also criticized because a few (but not all) investigators were unable to replicate his findings. Investigators who were unable to replicate, however, used a different method for stimulus presentation. Marcel used high-contrast cards and a tachistoscope and repeat experiments used low-contrast computers to present stimuli. A change in intensity of background between the original and repeat experiment could change the level of light (or dark) adaptation of the eyes and thus alter perception of stimuli.

Studies that used dichotic listening tasks were also criticized for not monitoring attentional control. It was argued that in these experiments volunteers were probably attending to the stimuli presented to both, the attended and unattended ear (or eye) but paid more attention to those on the attended side. Because of less attention paid to unattended stimuli, they may not have mentioned attending to them. This possibility undermines the significance of observations reported in these experiments.

Controvery about validity of data on nonconscious perception led to spirited debate on separability of perception, awareness, knowledge of awareness, and expression of knowledge. Since these functions are processed in separate brain areas they can be independently altered. It is therefore possible for a stimulus to be consciously perceived with full awareness but not registered in memory. Even if the stimulus is registered and the brain has knowledge of its existence, if the knowledge is not processed in the areas that are responsible for "expression" of knowledge (e.g., speech area), a volunteer may deny perception even after perceiving under full awareness. An interesting discussion on these issues is documented in an article published in 1986 by a Belgian psychologist Daniel Holender.[21] Holender was a critic of experiments on nonconscious perception and thus his paper lists studies that have not found evidence of learning from subliminally presented stimuli. The debate at the time was centered around a critical question - whether a stimulus can be cognitively processed without conscious perception.

To resolve this controversy, we set up a neuroimaging experiment. In this experiment, functional magnetic resonance imaging (fMRI) technique was used to study the pattern of brain activation elicited by subliminal stimuli in a memory task.[24] The objective of this experiment was to find out whether nonconsciously perceived stimuli are processed cognitively, and if so, if they use the same neural network that fully visible stimuli use. To elicit cognitive processing, healthy young volunteers were shown images of the common objects or animals for 3 seconds each and asked to memorize them (Fig. 1.5). Thereafter, both studied and novel images were shown subliminally for 27 ms. Volunteers were asked to indicate whether each sublimin image was studied, novel, or unrecognized. They were allowed to guess. Volunteers accurately recognized most studied images (77%) and did not recognize a majority (95%) of novel images presented subliminally. It was relatively easy for them to recognize studied images because those images were primed.

Analysis of the data revealed that subliminal images elicited increased activation in the prefrontal cortex bilaterally during presentation of studied subliminal pictures and unilaterally in the left hemisphere during presentation of novel subliminal pictures (Fig. 1.5). Both studied and novel pictures activated the hippocampus. These findings are significant because we observed the exact same pattern of activation when supraliminal, instead of subliminal, stimuli were used in a similar experiment conducted earlier.[25] In that study, increased activity of the left prefrontal cortex was associated with retrieval effort and successful retrieval elicited increased activity of the right prefrontal cortex. The fact that subliminal stimuli activated the areas that were active during similar processing of supraliminal stimuli suggests that the stimuli that are not consciously perceived are processed cognitively in the same brain areas that process consciously perceived stimuli.

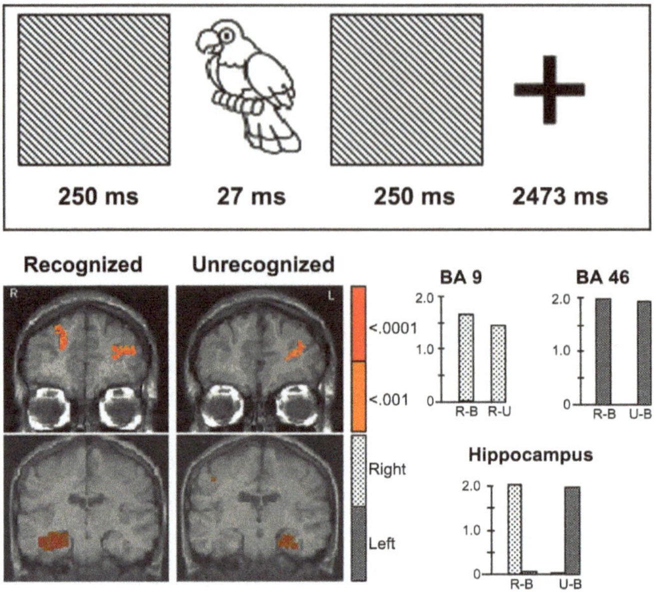

FIGURE 1.5 (Top) Subliminal stimuli were presented for 27 ms and both forward and backward masks were used to prevent afterimage. (Bottom) Subliminal stimuli that were primed, activated the prefrontal cortex of both hemisheres and the right hippocampus while unprimed and unrecognized subliminal stimuli activated the left hippocampus and left prefrontal cortex. The pattern of activation was similar to the pattern observed during processing of supraliminal stimuli. *Reproduced from Badgaiyan RD. Cortical activation elicited by unrecognized stimuli. Behav Brain Funct 2006;**2**(17):1–5.*

The most fascinating aspect of this experiment was activation of the prefrontal cortex and hippocampus during processing of novel subliminal stimuli that were not consciously recognized. Since volunteers were unable to recognize 95% of these images, they could not have made conscious attempt to match them with studied images (to decide whether it was drawn from the studied list). These images nonetheless activated the brain areas associated with retrieval effort—the left prefrontal cortex and hippocampus.[25,26] It appears that the brain tried to retrieve a match for stimuli that were not consciously perceived. This could not have happened consciously. How can we consciously try to find a match if the stimulus was not even perceived consciously? It thus appears that the brain did recognize those images nonconsciously but the recognition never reached conscious awareness.

In this experiment consciously recognized subliminal images were primed because of previous exposure at the study stage. Even though priming might have lowered the threshold for conscious perception, it probably did not affect cognitive processing—both recognized and unrecognized stimuli were processed exactly the way a supraliminal stimulus is processed under similar cognitive conditions. The only difference between subliminal and supraliminal stimuli was conscious awareness of recognition. It suggests that cognition and conscious awareness of a stimulus are processed in separate networks. Subliminal stimuli in this experiment did activate network for cognitive processing but not the network for conscious awareness. The two networks therefore are independent of each other. It also suggests that conscious awareness of a stimulus is not necessary for its cognitive processing. The awareness is processed in a separate neural network,[25,27,28] as discussed in Chapter 2, Nonconscious memory.

These results suggest that the brain receives and processes more information than we are consciously aware of, which means we do not know what our brain knows. We are not consciously aware of most of the information our brain retains in the nonconscious mind. We cannot access the informaiton retained in the nonconsiocus mind at will. This information however, is occasionally expressed in an unexpected way. For example, let us assume you do not like potatoes. If I ask why you do not like potatoes you probably will not have a credible explanation because it is a decision that was made by the nonconscious mind based on the "information" it has retained. It is possible that in the past you ate potatoes and had a bad experience. The nonconscious mind could have made a connection between the two events and triggered disliking for potatoes to avoid the "harmful event" in future. Similarly, while meeting a stranger we make an impression about his or her personality without knowing why we made that impression. We also consciously do not know why we feel comfortable in a place or in someone's company and uncomfortable in other situations. This means some of the information hidden in the brain helps us make decisions even though we are not consciously aware of the information we are using to make those decisions. These nonconscious decisions may or may not be the best or logical, as in the example of disliking potatoes, but these decisions are quick and avoid possible threat to life. Of course, we can decide to consciously override those decisions.

The nonconscious mind therefore tracks all experiences without our conscious knowledge. If we were able to access information retained in the nonconscious mind, our cognitive ability would be significantly enhanced. On the flip side, excessive information could confuse us and slow down the decision-making process. However. if we could acquire the ability to retrieve selected nonconscious information at will, we could expand our knowledge base and cognitive ability significantly. Additionally, at least theoretically, retrieval of nonconscious information may play a role in treating mental conditions. Sigmund Freud's (Fig. 1.6) theory of the subconscious mind is based on precisely this idea.[5,29] Freud suggested that unresolved conflicts in nonconsciously retained information alter our cognition and behavior leading to mental illness. By bringing those conflicts to conscious mind, it may be possible to work on resolving the conflict. The resolution could potentially

FIGURE 1.6 Sigmund Freud (1856–1939). Proposed an influential theory on the unconscious mind and its role in pathophysiology of mental illnesses. *Reproduced from Wikimedia Commons.*

reduce some psychiatric symptoms. Even though Freud's concept of nonconscious (which he called unconscious) is not exactly what we are discussing here, but the idea is similar. His unconscious mind retains information from past events and does not make them available to conscious awareness. Freud's psychotherapeutic techniques to reveal information retained in the nonconscious mind are still used today with some modification.

Psychotherapy is not the only method developed to bring nonconscious information to conscious awareness. It is claimed that the information can be retrieved under hypnosis and under influence of drugs and alcohol. Indeed, in the last century people used a variety of drugs in an attempt to reveal inner conflicts. Even today, some of these techniques are occasionally used for forensic purposes to unravel the truth! The most popular drug used for this purpose is a barbiturate, amytal sodium (amobarbital sodium).

As you might guess, none of these techniques have proven to retrieve hidden nonconscious information. But we do know that some of this information is retrieved involuntarily under certain conditions, but we have no voluntary control over the retrieval, or for that matter, over selection of the information retrieved. The nonconscious mind has its own "mind" and determines which information to release and which to keep hidden from conscious awareness. It is an area that clearly needs further research.

If we are able to retrieve nonconscious information at will, we could use it to treat at least some of the mental conditions like amnesia. Amnesic individuals lose their conscious memories and forget their names, addresses, and other identifying information. To bring their memory back, an attempt is made to retrieve nonconscious autographical memory. One such technique was originally developed by a British psychologist Sir Francis Galton (Fig. 1.7) in 1879.[30] His technique was revived about a century later by two Duke University psychologists, Herbert F. Crovitz and Harold Schiffman.[31] In this technique, an amnesic individual is asked to free associate with a cue word or say the first word that comes to mind when a cue word is presented. Nonconscious association with the cue provides information to help establish identity. For example, if an individual is asked to say the first word that comes to mind with the cue word "food" and if the answer is "pizza," and if

FIGURE 1.7 Francis Galton (1822–1911). Credited with inventing a technique to retrieve nonconscious information in amnesic patients. *Reproduced from Wikimedia Commons.*

he names a pizza store in response to the next question, he is giving clue about his or her work place or a favorite joint. This information could be further narrowed down by having the individual free associate with names of cities, streets etc. However, this technique has limited use because it does not always retrieve desired information. If we could acquire the ability to retrieve nonconscious information at will, it would be easier for amnesic individuals to recall autobiographical information. It could help not only amnesic individuals but people with other conditions as well.

This brief history of the science of nonconscious mind will help us understand specific nonconscious processes discussed in the following chapters.

Bibliography

1. Patañjali, Dvivedi MN, Subrahmanya Sastri S. *The yoga-sūtras of Patañjali*. Theosophical Publishing House; 1934.
2. Radhakrishnan S, Moore CA. *A source book in Indian philosophy*. Princeton University Press; 1957.
3. Mehta J. *Essence of maharishi Patanjali's Ashtang yoga*. Pustak Mahal; 2005.
4. Hartmann EV, Coupland WC. *Philosophy of the unconscious; speculative results according to the inductive method of physical science*. new ed. K. Paul, Trench Harcourt, Brace & Company; 1931.
5. Freud S. The unconscious. In: Strachey J, editor. *The standard edition of the complete psychological works of Sigmund Freud*, vol. 14. Hogarth Press and Institute of Psycho-analysis; 1915. p. 159–215.
6. Peirce CE, Jastrow J. On small differences in sensation. *Memoires Natl Acad Sci* 1884;**3**:73–83.
7. Sidis B. *The psychology of suggestion*. D. Appleton and Company; 1898.
8. Breant H. "Hidden sell" technique is almost here. *Life* 1958;**44**:104.
9. Key WB. *Subliminal seduction; Ad media's manipulation of a not so innocent America*. Prentice-Hall; 1973.
10. Greenwald AG, Spangenberg ER, Pratkanis AR, Eskenazi J. Double-blind tests of subliminal self-help audiotapes. *Psychol Sci* 1991;**2**(2):119–22.
11. Vokey JR, Read JD. Subliminal messages. Between the devil and the media. *Am Psychol* 1985;**40**(11):1231–9.

12. Baker LE. The influence of subliminal stimuli on verbal behavior. *J Exp Psychol* 1937;**20**:84–100.
13. Dixon NF. The conscious-unconscious interface: contributions to an understanding. *Arch Psychol (Frankfurt)* 1983;**135**(1):55–66.
14. Lewis JL. Semantic processing of unattended messages using dichotic listening. *J Exp Psychol* 1970;**85**(2):225–8.
15. Walker P. Orientation-selective inhibition and binocular rivalry. *Perception* 1978;**7**(2):207–14.
16. Greenwald AG. New look 3: unconscious cognition reclaimed. *Am Psychol* 1992;**47**(6):766–79.
17. Marcel AJ. Conscious and unconscious perception: an approach to the relations between phenomenal experience and perceptual processes. *Cognit Psychol* 1983;**15**(2):238–300.
18. Marcel AJ. Conscious and unconscious perception: experiments on visual masking and word recognition. *Cognit Psychol* 1983;**15**(2):197–237.
19. Bruner JS, Postman L. Tension and tension release as organizing factors in perception. *J Pers* 1947;**15**(4):300–8.
20. Erdelyi MH. A new look at the new look: perceptual defense and vigilance. *Psychol Rev* 1974;**81**(1):1–25.
21. Holender D. Semantic activation without conscious identification in dichotic listening, parafoveal vision and visual masking: a survey and appraisal. *Behav Brain Sci* 1986;**9**:1–23.
22. Green DM, Swets JA. *Signal detection theory and psychophysics*. R. E. Krieger Pub. Co; 1974.
23. Cheesman J, Merikle PM. Priming with and without awareness. *Percept Psychophys* 1984;**36**(4):387–95.
24. Badgaiyan RD. Cortical activation elicited by unrecognized stimuli. *Behav Brain Funct* 2006;**2**(17):1–5.
25. Badgaiyan RD, Posner MI. Time course of cortical activations in implicit and explicit recall. *J Neurosci* 1997;**17**(12):4904–13.
26. Badgaiyan RD. Neuroanatomical organization of perceptual memory: an fMRI study of picture priming. *Hum Brain Mapp* 2000;**10**(4):197–203.
27. Badgaiyan RD, Posner MI. Priming reduces input activity in right posterior cortex during stem completion. *Neuroreport* 1996;**7**(18):2975–8.
28. Badgaiyan RD. Conscious awareness of retrieval: an exploration of the cortical connectivity. *Int J Psychophysiol* 2005;**55**(2):257–62.
29. Freud S. *New introductory lectures on psychoanalysis*. New York: Carlton House; 1933.
30. Galton F. Psychometric experiments. *Brain* 1879;**2**:149–62.
31. Crovitz HF, Schiffman H. Frequency of episodic memories as a function of their age. *Bull Psychon Soc* 1974;**4**:517–18.

Chapter 2

Nonconscious memory

The concept of nonconscious learning and memory is not new. Incidental learning, which essentially means encoding of memory without conscious awareness, is known for decades. Over 100 years ago in 1913 a New York psychologist named Garry Cleveland Myers (1884–1971) published a 108-page supplement in *Archives of Psychology*.[1] It included over 100 experiments on memory without awareness. Myers is better known for publishing the children's magazine *Highlights for Children* along with his wife Caroline. There is a personal tragic story associated with this magazine. Highlights for Children was popular in Pennsylvania but had no national circulation. To expand its circulation Myers sent his son Garry Jr. and wife Mary to New York for distribution in that area. On the way, both of them died in a mid-air collision between TWA flight 266 from Dayton, OH Ohio and United flight 826 from Chicago, IL Illinois over Staten Island on December 16, 1960. It was the worst air disaster in US history at the time. A total of 128 people died.

A New York psychologist named James McKean Cattell (Fig. 2.1) was credited with conducting the first experiment on incidental learning, and was also responsible, in part, for establishing psychology as a scientific discipline. Cattell conducted the first experiment on incidental learning in one of psychology classes at Columbia in 1893 and published the data in 1895.[2] He designed the experiment after a fierce debate with Harvard psychologist William James over medium Leonora Piper. Cattell believed that we are not consciously aware of all behavioral and cognitive processing. Since James disagreed, Cattell conducted an experiment to prove his point. He asked students to answer a series of questions and to rate how confident they were in the accuracy of their answers. Questions were carefully crafted to ensure that answers could have been learned only incidentally. For example, how much time does it take to walk from the college entrance to the lecture hall? What is the weight of their textbook (William James' *A Briefer Course in Psychology*)? What was discussed in the first two minutes in the lecture held a week ago? He found that even though students had little or no confidence, they were able to answer those questions accurately. This discrepancy between accuracy and level of confidence suggested that the students had no conscious knowledge of the source of their answers. Cattell proved that we

Neuroscience of the Nonconscious Mind. DOI: https://doi.org/10.1016/B978-0-12-816115-9.00002-4

FIGURE 2.1 James McKean Cattell (1860–1944) credited with conducting the first experiment on nonconscious learning. *Reproduced from Wikimedia Commons.*

are not consciously aware of all the information our brain retains and all cognitive processing it carries out. This experiment created history by initiating research on incidental learning and memory.

Around the same time a French scientist named Alfred Binet (Fig. 2.2) made a significant discovery concerning incidental learning while working with chess players and school children. Binet studied law and natural science and is best known for developing the first IQ test (Binet–Simon Test). Even though he had no formal education in psychology, he was the head of the Laboratory of Experimental Psychology at Sorbonne France from 1894 until his death in 1911. He conducted several ground-breaking experiments. In a highly cited experiment, Binet asked experienced and novice chess players to play chess blindfolded. He found that the experienced players were able to play multiple boards simultaneously and accurately remembered "the game" in all boards. While they had no photographic memory of boards and did not know the positions of different pieces, the players were able to reconstruct the whole game. In Binet's opinion, these players were using a form of memory that is different from the usual (conscious) memory known to psychologists at the time. He published the results in 1894 in a book titled *Mnemonic Virtuosity: A Study of Chess Players.*[3] He also studied child prodigies who could do math on 21-digit long numbers at age six. Based on the results of the experiments he conducted on school children and on his own daughters, Binet concluded that there are different forms of memory, and some of them are not under conscious control. Based on this finding, in the IQ test he incorporated questions that are dependent on incidental learning.

FIGURE 2.2 Alfred Binet (1857–1911). Developed the IQ test and studied incidental learning. *Reproduced from Wikimedia Commons.*

Unfortunately, Binet's contribution to science was not recognized in his own country while he was alive. However, his intelligence test became popular outside France, and experiments on chess players and schoolchildren prompted other investigators to study incidental memory. These experiments were documented in Myers' work.[1] In most of these experiments, volunteers were asked to remember letters, and numbers, or to estimate elapsed time, or the size and weight of various objects. At that time, experimental paradigms were not standardized, and investigators reported large variation in the results. The general belief at the time was that intentional (conscious) memory is superior to incidental (nonconscious) memory but not all psychologists believed this. One of them, Sadie Rae Myers Shellow, a professor at Columbia University, wrote that "incidental memory is only capricious if we do not know its determining factors and once these are analyzed and the laws governing them formulated by experimental research, we will find incidental memory as logical and as inevitable as the better accredited direct memory."[4] These views were clearly ahead of her time. She thought that events retained in incidental memory were those "which had become fixated because of their own intensity, was sufficient to cause fixation, or because they had in some way become linked up with the individual's habitual interests and past experience and had become incorporated with them."[4]

Despite this forward-looking view, most investigators at the time continued to believe that incidental learning was inferior to intentional learning.[5] As a result, enthusiasm to study the phenomenon waned for a long time. Interest was revived in 1967 after publication of a seminal paper by an American psychologist Arthur S. Rebar (born 1940) who worked at Brown University in Rhode Island. Influenced by contemporary debate on language learning, he used a linguistic approach to study what he called implicit learning. In one experiment, he presented a list of 28 strings of letters (each consisting of 3–8 letters) and then asked volunteers to memorize the strings. The strings did not make a legitimate word (e.g., BTSXSE and BPTT) but volunteers were told that they follow rules of an artificial grammar. Next, he presented 88 strings and asked volunteers to indicate whether those strings followed the same rules of the artificial grammar. The strings indeed followed rules of the finite-state grammar developed by linguist Noam Chomsky and his colleagues at MIT. It had 18 rules. In Reber's experiments, even though none of the volunteers consciously understood those rules, their responses (69.4% correct) were significantly better than chance.[6] This indicated that they had learned the rules nonconsciously. After this observation was replicated by other investigators, there was renewed interest in the study of implicit learning. Reber's study debunked the notion developed by Chomsky and other psycholinguists that the brain has a special learning module for language. Reber showed that language is learned using an implicit learning mechanism that is same for linguistic and non-linguistic learning. He noted that this allows us to recognize a pattern in our environment to effectively understand and respond to it. He wrote, "axiomatic and conscious cognitive processing cannot take place without using implicit learning." This statement turned out to be true.

Even though the results of his experiment on artificial grammar were replicated by many investigators there were lingering doubts about implicit learning. It was argued that while in these experiments volunteers did not learn all 18 rules of the grammar, they may have noticed a few rules that could have allowed them to perform at above chance level. This criticism led investigators to study implicit learning during sleep, hypnosis, and under general anesthesia. These experiments initially suggested that instructions given during sleep or under hypnosis are indeed followed, indicating that memory of information received under these conditions is retained.[7,8] However, these results were not considered reliable because there was no control and observers were not blinded. Studies that were well designed did not find evidence of either explicit or implicit learning during sleep.[9] Even though there is no proof that learning happens during sleep, it is known that memory of learned events is recalled nonconsciously during sleep. For example, people with somnambulism or who sleep -walk, can perform complex tasks without conscious awareness while still sleeping.

The results of experiments on anesthetized individuals are interesting. In one experiment[10] individuals under the general anesthesia isoflurane were presented pairs of semantically related words (e.g., doctor and nurse could be one such pair). After recovery, they were tested for conscious (explicit) and nonconscious (implicit) memory. They had no explicit memory of studying word pairs and were unable to recall the other word of a pair when one word was presented as a cue. However, when asked to say the first word that came to mind after being given a cue word that was one member of the pair, the subjects often responded with the other word of the pair. This effect is a form of nonconscious memory called priming. A metaanalysis[11] of 12 studies involving 708 anesthetized patients concluded that there was a highly significant priming effect when memory was tested within 12 hours following administration of isoflurane anesthesia. However, these results did not replicate when the anesthetic agent was changed to sufentanyl.[12]

It is not clear why different anesthetic agents have different effects. It could be due to the complexity of modern anesthetic techniques. These techniques require individuals to receive a combination of drugs for analgesia, muscle relaxation, and prevention of awareness of the procedure. While these techniques are almost always effective, a few patients (0.13%) are known to retain at least partial awareness of the procedure while under anesthesia.

Since it is difficult to estimate how many patients remain conscious under anesthesia, it is unclear whether the priming effect observed in anesthetized individuals is due to retained awareness during the procedure. This possibility was examined by continuously recording brain activity or electroencephalogram (EEG) in the operating room. Based on the activity, power of the EEG waves is estimated. Based on this, a score called the bispectral index score (BIS) is calculated. It has a range of 0 (minimal brain activity) to 100 (fully awake). A score between 40 and 60 is considered an optimal anesthetic score. Using BIS as a measure of awareness, Smith and Maye[13] studied semantic memory in 19 surgical patients with different BIS to examine whether retention of knowledge under anesthesia depends on the level of awareness. They did not find a significant correlation between BIS and memory performance, suggesting that learning during anesthesia may not be due to retained awareness.

Learning and memory under sleep and anesthesia however are not the same as other nonconscious processes and strictly speaking they do not meet the criteria I have laid out in the beginning for nonconscious process. I defined nonconscious process as the one that is accomplished while a person is fully awake and responsive, but under anesthesia and sleep an individual remains unresponsive and unaware of the surrounding. Thus, brain processing of stimuli under these conditions may be different from that observed during nonconscious processing when fully awake. Study of nonconscious processes during sleep and anesthesia is important to understand the nature of learning and memory under these cinditions, however.

Understanding the brain mechanisms that retain, retrieve, and process nonconscious information requires careful observation and diligent interpretation of data. In early experiments, validation of experimental findings was ignored because of the excitement over unraveling "hidden knowledge". It lead to decades of uncertainty and skepticism in this area of study. One such experiment was conducted by the famous neuroscientist, Wilder G. Penfield (Fig. 2.3), who was born in the United States (Spokane, WA) but accepted Canadian citizenship after he was not allowed to establish a center to study epilepsy using an endowment from David Rockefeller. He used the endowment to establish the Montreal Neurological Institute at McGill University.

As a standard treatment of epilepsy, Penfield used to remove parts of the brain identified as epileptic foci. He thus had the opportunity to study different areas of the brain. Using this opportunity, he meticulously mapped motor and sensory areas of the brain and made the famous homunculi maps that all students of the brain are familiar with (Fig. 2.4). He also made significant contribution to our understanding of hallucination and the *déjà vu* phenomenon. Interestingly, he began his research career by searching seats of consciousness and soul in the brain. He did not find either, but he made significant discoveries that have helped us understand the functions of the brain.

FIGURE 2.3 Wilder G. Penfield (1891–1971). Rediscovered the role of the temporal lobe in memory processing. *Reproduced from Wikimedia Commons.*

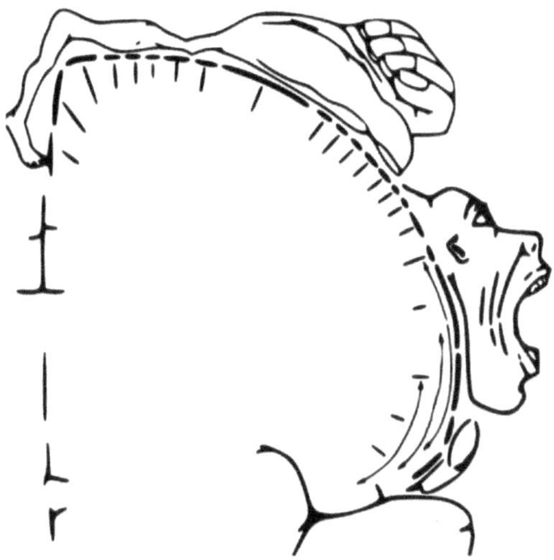

FIGURE 2.4 Cartoon of the section of brain showing the motor homunculus representing areas of the motor cortex that control different parts of human body. *Reproduced from Wikimedia Commons.*

Penfield conducted several experiments to study registration and recall of memory. He presented the results of those experiments in 1951 as a part of the 76th annual meeting of the American Neurological Association in Atlantic City, NJ and published them in 1952.[14] His most fascinating and influential experiment was conducted on an epileptic individual during a neurosurgical procedure. Penfield reported that "when the first convolution of right temporal lobe was stimulated, the patient began saying there is a piano here and I can hear the song, you know." This statement was interpreted as recollection of a past memory. It was extremely vivid and appeared as if he was there at that time describing the event as it was occuring. Interestingly, this individual had no recollection of "the event" when stimulation was terminated. The observation at the time was revolutionary because it suggested that vivid memories of all of our experiences are retained in the brain even though we cannot recall them. The patient's "recollection" was interpreted as memory he could not consciously access. Penfield suggested that "the memory record continues intact even after the subject's ability to recall disappears."

Penfield's finding was popularized by the book *I'm OK, You're OK* published by Thomas Anthony Harris in 1969.[15] The book was translated into 12 languages and sold over 15 million copies. It is a practical self-help guide to solving emotional problems based on a psychoanalytic theory developed by the American psychiatrist Eric Berne. Since Penfield's experiments were described in details in the book, popularity of his experiments rose along with that of the book.

Despite their popularity, Penfield's experiments were not perfect. There were many unanswered questions. For example, he conducted experimentsonly on epileptic individuals known to have an "aura" that includes sensory experience. It is unclear whether his patients' "recollections" were a form of aura or actual recall of a past event. Penfield himself appears to be unsure about the nature of recalled experience because as he said, "In these experiments it seems to make little difference whether the original experience was fact, dream or fancy." He never confirmed whether "recollected" events actually happened. Moreover, the "recollection" occurred in only 5% of epileptic individuals. If this were the normal brain function, one would expect it to happen more often. The credibility of his results was further eroded by the observation that the "recollection" did not go away when part of the temporal lobe that elicited those "memories" was surgically removed. Penfield explained this by suggesting bilateral "storage" of information. Indeed, bilateral damage to temporal lobe causes serious memory deficits. It is still unclear whether Penfield's observations were vivid recollections of past events or some form of hallucination, aura, or dream-like replay of an imaginary event.

Temporal lobe damage and the story of HM

Even though no clear inference could be drawn from Penfield's experiments, his observation rekindled interest in the study of the temporal lobe and its role in memory processing. The temporal lobe was first associated with memory by a Russian psychiatrist named Vladimir Mikhailovich Bekhterev (Fig. 2.5)[16] who observed "softening" of the medial temporal lobe (MTL) in the autopsied brain of an amnesic patient. Unfortunately, this finding was not followed up on in part due to political reasons. Bekhterev graduated from the Russian Military Academy and became Chair of Psychiatry when he was only 28 years old. Eight years later he moved to St. Petersburg as Professor and Director of Psychiatry at the Military Medical Academy. He authored over 1000 scientific papers and created at least two new disciplines: reflexology and psychoneurology. He described reflexology as the study of response to external or internal stimuli and psychoneurology as the study of disease based not only on anatomy and physiology but also on philosophical, psychological, and social issues. He therefore pioneered the biopsychosocial model of disease we know today. But he was at odds with the Russian communist regime and died mysteriously in 1927. After his death, Soviet dictator Joseph Stalin ordered removal of his name and all references to his research from books and scientific literature.

Since Bekhterev's work did not disseminate, his observation of the association between the temporal lobe and memory was forgotten until Penfield rediscovered it and began studying people with temporal lobe lesions. After studying one such individual in Canada, his group found another person in

FIGURE 2.5 Vladimir Bekhterev (1857–1927). First made the connection between the temporal lobe and memory. *Reproduced from Wikimedia Commons.*

FIGURE 2.6 Henry Gustav Molaison (1926–2008) was the famous patient, HM, who was the subject of over 100 studies on conscious and nonconscious memory. *Reproduced from Wikimedia Commons.*

the United States. This individual, widely known as HM in the literature, would go on to become the most studied patient in medical literature.[17] HM was born Henry Gustav Molaison (Fig. 2.6) in Hartford Connecticut in 1926. After a bicycle accident at age 7, he started having epileptic fits (first partial seizures and then tonic-clonic seizures at age 16). He graduated high school and started working as a motor winder but could not keep the job for long because of frequent epileptic attacks. Since medications did not control his seizures he was referred to neurosurgeon William Beecher Scoville at Hartford Hospital. Scoville followed Penfield's lead and surgically removed the part of the brain that generated the seizures. On August 25, 1953, he removed HM's medial temporal lobe (MTL) that included the anterior two-thirds of the hippocampus, the parahippocampus, the entorhinal cortex, the piriform cortex, and the amygdala of both hemispheres. Part of his anterolateral temporal cortex was also removed. Initially about two centimeters of the hippocampus was left intact, but later it was atrophied and lost all connections to or from other parts of the brain. HM's surgery was uneventful, but he developed amnesia after the procedure. Between 1957 and before his death in 2008, HM was studied by many investigators and more than 100 papers were written about him. After death, HM's brain was preserved and housed at the University of California San Diego. In 2009 it was sliced into histological sections and in 2014 a 3D digital reconstruction was made available to investigators. His brain is still being examined extensively with funding from the Dana Foundation and National Science Foundation.

At the time HM was having surgery, Brenda Milner (Fig. 2.7), a researcher in Penfield's laboratory at McGill University, was looking for a patient with a bilateral temporal lobectomy. Milner was born in Manchester, England and graduated with a degree in experimental psychology from

FIGURE 2.7 Brenda Milner (born 1918). Defined the role of temporal lobe in memory processing. *Reproduced from Wikimedia Commons.*

Cambridge University. During World War II she was tasked to use aptitude test to distinguish between fighter and bomber pilots. While working for the military she met her husband who decided to move to Canada to take up a position in an atomic research laboratory. Milner moved along and started working on her Ph.D. thesis at McGill University under supervision of a famous psychologist Donald Hebb. While working on her thesis she became interested in a patient nicknamed PB who had his temporal lobes removed. PB had an interesting history. Initially the temporal lobe of only one side of his brain was removed but later that of the other side was also removed. Removal on one side did not make much of a difference but after it was removed on both sides, he became amnesic. To confirm this finding, Milner needed another person with bilateral temporal lobectomy to examine Penfield's hypothesis suggesting that memory is processed in the temporal lobe.[18] When Milner learned about HM she contacted Hartford surgeon Scoville and studied HM for the next several years. This study helped her make important discoveries about the role of MTL in memory processing. She recently celebrated her 100th birthday and as of the date of publication still works as a Professor of Neurology and Neurosurgery at McGill University and as a Professor of Psychology at Montreal Neurological Institute.

Very early in her interaction with HM, Milner found that he remembered events prior to the surgery but could not remember recent events. For example, once he had breakfast with Milner and within 30 minutes, not only forgot what he had for breakfast but even forgot having breakfast. HM used to say philosophically, "every day is alone in itself, whatever enjoyment I have had and whatever sorrow I have had." So true if one loses memory. Memories maintain continuity of life events.

Milner soon found that despite amnesia HM was capable of learning and remembering certain tasks. In 1959 he performed 59 trials of a mirror-drawing task in which a volunteer traces a drawing reflected on a mirror. Time for completion and number of errors is recorded. It is a nonconscious task of motor learning. HM showed steady improvement in response time over 3 days,[19,20] even though he did not remember completing the task on previous days. In 1962 HM was asked to learn to correct the sequence of turns in a maze. Even after completing several trials he was unable to recall the sequence, but his response time improved significantly after practice. These experiments suggested that despite the MTL lesion, HM was able to nonconsciously learn new tasks.[21] These tests were conducted by one of Brenda Milner's students, Suzanne Corkin (Fig. 2.8), who later became a Professor in the Department of Brain and Cognitive Sciences at MIT. Encouraged by these findings, Corkin examined HM's performance on other tasks that required nonconscious learning of motor movements. He was able to learn and retain information in most of those tasks for several days.[22,23] In 1966 HM was admitted to the Clinical Research Center at MIT for 2 weeks

FIGURE 2.8 Suzanne Corkin (1937–2016) studied the memory of HM. *Reproduced from Wikimedia Commons.*

for further in-depth study. At that time, he had no difficulty remembering events that happened and people he met prior to the surgery but was unable to remember postsurgical events and the names of people he met after the operation. He used to speak in monotone and had lost interest in sex. He rarely asked for food and had normal comprehension of language. After arriving at the clinic, he did not remember where he was and how he got there. He asked the nurse at least three times every day where he was and how he got there. At that time, his mother was in an inpatient unit of Hartford Hospital for a minor surgery. Even though HM visited her three times in the hospital, he had no recollection of her being there. However, he felt uneasy several times and thought one of his parents was going through a difficult time.

At the Clinical Research Center, his cognitive abilities were examined using a battery of tests. It was observed that despite severe amnesia HM's performance was surprisingly better than those of healthy control volunteers on Wisconsin Card Sorting Task, which requires volunteers to match cards with similar features. Since cards are designed to have many features, the match requires meticulous planning and shifting strategies. His performance was also better than healthy control volunteers on Mooney's face-perception test, which requires estimation of the age and sex of fragmented pictures of human faces (Fig. 2.9). But in tests of conscious memory, HM had

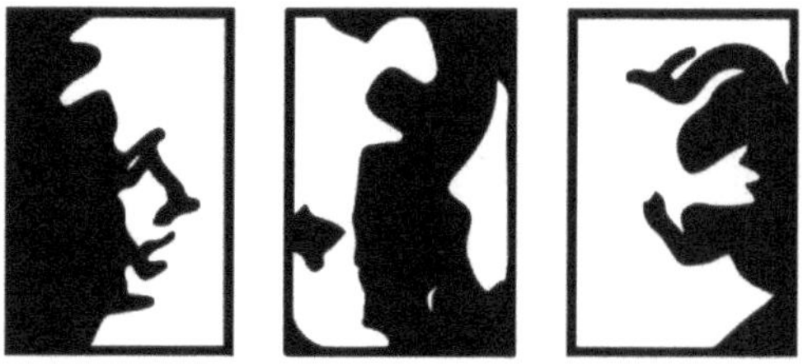

FIGURE 2.9 Example of Mooney's face-perception test. In this test HM performed better than healthy control volunteers. *Reprinted from Verhallen RJ, Bosten JM, Goodbourn PT, Bargary G, Lawrance-Owen AJ, Mollon JD. An online version of the Mooney Face Test: phenotypic and genetic associations. Neuropsychologia 2014;**63**:19–24.*

significant impairment. Even though his attention span was normal, he could consciously retain three-digit numbers for about 15 minutes.[22]

It appears that HM gradually lost his autobiographical memory. When first examined in the 1950s, he had memory of events that happened before he was 24 years old.[24] In 1985 he could remember events that happened when he was 16 years and younger,[25] and at age 76 he could not recollect any event specific to time and place.[26] At that age, he also had cortical thinning and other age-related changes in the brain, which Corkin did not reveal at the time.

Soon after her death in 2016, Corkin was accused of suppressing facts by a journalist (Luke Dittrich). She was also accused of keeping HM's identity a secret and experimenting with him without obtaining consent from his close relatives. She was even accused of shredding results of some of her experiments on HM. The journalist apparently had interviewed Corkin shortly before her death, while she was sick and on chemotherapy for liver cancer. The journalist wanted Corkin's work on HM to be retracted, yet her institution, MIT, strongly refuted all allegations.

Corkin and Milner's work on HM established that even though there was almost a complete loss of conscious memory, some form of memory is retained in people who have bilateral damage to the MTL. Additionally, those retained memories are not consciously available but are expressed while performing tasks like maze and mirror drawing. HM also retained memory of tasks that we frequently repeat, like conversations, walking etc. All of these tasks involve nonconscious mind and are examples of nonconscious memory.

At the time, Milner and Corkin were experimenting on HM, other researchers were studying factors that affect recollection. One of them was Lowell H. Storm at the University of Minnesota. While working on his PhD dissertation in 1956, Storm found that recollection of words shown to volunteers improved significantly if the word is presented after they had studied a related word.[27,28] For example, if the word "bird" was studied in an unrelated context, it was easier for volunteers to subsequently recall the word "eagle," even though the word bird was not supposed to be remembered and volunteers did not remember studying it. This observation indicated that mere mention of the word bird primed semantically related words. Storm called this phenomenon the "recency effect," but it is now referred to as semantic priming. The term "priming" was first used by Charles Cofer of University of Maryland[29] and Phebe Cramer of University of California, Berkley.[30] They independently developed paradigms similar to the one developed by Storm, around the same time.

This semantic priming task was refined, and several variations were developed in the early 1970s. A popular variation called the lexical decision task was developed by David Meyer of Bell Telephone Laboratories in New Jersey and Roger Schvaneveldt at the University of Colorado. Together, in 1971 they published a seminal article on semantic priming.[31] In the lexical decision task, a volunteer is shown a prime word followed by either a legal word or a string of pronounceable nonwords (e.g., WIXET). The task requires them to decide whether the second stimulus is a word or nonword, as quickly and as accurately as possible. It was found that if the prime word was semantically related to the second stimulus the volunteer's response was quicker. Thus, response to the word "doctor" would be significantly quicker if it was preceded by the word "nurse" rather than "nun." This form of semantic priming was first studied at Stanford University by Thomas Landauer and Jonathan Freedmas.[32] It was further developed by Don Scarborough and colleagues at the City University of New York.[33] Researchers have now developed many tests that elicit semantic priming. In all of these tests a stimulus presented as a primer is processed nonconsciously to improve the response of a consciously performed cognitive task.

Semantic priming involves processing of stimuli for its meaning, and many researchers believe it is not a real nonconscious task because it probably involves components of conscious processing. It does not necessarily mean a volunteer is consciously aware of the processing. It only means that it involves some of the processing that is normally activated during conscious recall. This argument suggests that priming tasks that are not dependent on the meaning of a stimulus are real nonconscious tasks. This condition is met in another form of priming called perceptual priming, in which prime and target stimuli have perceptual similarity. An example of this form of priming is the incomplete picture priming task, with which HM was also tested.[22,34] In the task, animals and common objects are drawn at different levels of degradation. Volunteers are first shown the most degraded picture and asked to identify the object. The process is repeated with clearer pictures until the object is recognized (Fig. 2.10). Number of errors made before correct recognition is counted to score the performance. The procedure is repeated after several hours or days. Because of the priming effect, performance improves in subsequent sessions and people make fewer errors.

FIGURE 2.10 Examples of fragmented images used in incomplete picture priming tasks. *Adapted from Snodgrass JG, Feenan K. Priming effects in picture fragment completion: support for the perceptual closure hypothesis. J Exp Psychol (General) 1990;**119**(3):276–296.*

In addition to semantic and perceptual priming tasks, there are other tasks that involve both semantic and perceptual priming depending on the experimental protocol used. One of those tasks is the word-stem completion (WSC) task, which was first used by British psychologists Elizabeth Warrington and Lawrence Weiskrantz (Fig 3.1) in 1970.[35] In this task the volunteer is briefly shown a list of words called the study list. During the study, they are given another task such as, counting vowels or deciding whether they like or dislike a word. This strategy ensures that each word is attended to. It also prevents conscious attempts to memorize words. After the list is studied, the first three letters of a word called a word stem is presented and the volunteer is asked to say aloud the first word that comes to mind beginning with the stem as quickly and as accurately as possible. Even though each stem makes multiple words, volunteers tend to complete word stems using a studied word, while consciously remaining unaware of the fact that the retrieved word was studied previously. Thus, if the list had the word 'picnic', a volunteer would most likely complete the word stem 'pic' with the word picnic rather than picture or pickup. For this protocol, it is important to ensure that no two study words have the same word stem and that each stem makes multiple (at least six) commonly used legal words. In this task, stems of unstudied words are also presented along with those of the studied words. The completion rate of the stems of unstudied words is used as a control to estimate priming effect. This task can be used to elicit either perceptual or conceptual priming depending on the design. If there is perceptual similarity between study items and stems, the task uses the perceptual priming mechanism, but when there is no such similarity, it uses the conceptual priming mechanism as discussed later in this chapter. To elicit conscious or explicit memory in this task, volunteers are instructed at the study stage to memorize words for later retrieval, and during word-stem presentation, are asked to complete each stem using a studied word.'[36–38]

The demonstration that nonconsciously studied words affect subsequent retrieval opened up a new chapter in memory research. Based on these and other findings, in 1985 Peter Graf and Daniel L. Schacter (Fig. 2.11) of the University of Toronto (Schacter is currently the Chairman of Psychology Department at Harvard University) defined two forms of memory: implicit or nonconscious and explicit or conscious.[39] Implicit memory includes all forms of conceptual and perceptual priming in addition to procedural or motor memory. All forms of memory that are performed without conscious awareness come under the broad umbrella of implicit or nonconscious

FIGURE 2.11 Daniel L. Schacter (born 1953) made significant contribution to our understanding of implicit and explicit memory. *Image courtesy of Prof. Schacter.*

memory. Motor or procedural memory refer to learning of sequence of muscles movements. Thus, if we repeatedly move hands, fingers, or any other part of the body in the same sequence, the sequence is learned and when needed, it is nonconsciously repeated while we remain consciously unaware of the sequence. A good example of this type of learning is typing. The fingers of an expert typist move to the appropriate locations on the keyboard without much conscious effort because the movement required to push a specific key is retrieved nonconsciously. Walking and negotiating staircases are other examples of procedural memory.

After it was found that HM had intact procedural memory while being densely amnesic, his ability to process other forms of implicit memory including priming was studied using a variety of tasks. Perceptual priming was studied using the incomplete picture task described above. He was examined twice on this task, once by Milner[22] and again by then graduate student of Suzanne Corkin and currently an MIT professor John Gabrieli.[40] In both studies HM made fewer errors in recognizing fragmented pictures in repeat presentations compared to the first presentation. When fragmented pictures were shown for the very first time by Milner, HM made 21 errors, but when they were presented for a second time, he made only 11 errors, suggesting nonconscious learning and retrieval. In Gabrieli's experiment, average number of errors was 22.4 and 14.8 in the first and second presentations respectively. Improvement in the second presentation was significant particularly because HM did not consciously remember that he was shown pictures in the first presentation. He thought he was taking the test for the very first time. These experiments indicate that he learned the skill nonconsciously and had intact perceptual priming.

HM's performance on tests of conceptual priming using lexical decision tasks (described above) was interesting.[41] It was comparable to that of healthy control volunteers for words that were used before the surgery but was significantly impaired for words included in English language (new dictionary entry) after his surgery. Since he was unable to form new memories after the MTL removal, his knowledgebase was limited to presurgery years. For that knowledgebase, his performance was not impaired. It suggests that the task of conceptual priming is dependent on the conscious memory mechanism as discussed earlier, and that the conscious and nonconscious (priming) memory systems are interdependent.

This interdependence was observed in another experiment in which HM was tested by John Gabrieli in WSC task. In this experiment HM completed stems using only those words that he had known before the surgery. Thus, when word COMPLAINT was shown, both HM and normal volunteers completed the stem COM using the word complaint. However, for word FRISBEE, HM completed the stem FRI using the word Friday because the word frisbee did not come into common use when HM was operated upon. He thus exhibited normal nonconscious learning for words familiar to him preoperatively but not for the words he had not learned before the surgery. This observation also suggests how explicit memory affects implicit memory retrieval.

Additionally, HM had intact stimulus-response learning under both classical[42] and instrumental conditioning.[43] Classical conditioning was studied using eye-blink conditioning. In this study HM was asked to listen to an auditory tone and relax. Immediately after (400 ms) the tone, a puff of air was delivered in the eyes causing eye blink. After pairing the tone and puff repeatedly, HM learned to blink on hearing the tone, before air puffs were delivered. He was tested again on this task two years later and this time he developed eye blink conditioning in one-tenth of the original time, even though he had no conscious recollection of performing the task earlier. He therefore was able to learn nonconsciously using classical conditioning paradigm.

A variant of classical conditioning is operant conditioning, which tests the ability to learn by analyzing the consequences of a behavior or action. Thus, if an action rewards, we nonconsciously learn to repeat it and if punished we learn to avoid the action. It has survival benefits and because of this, we learn throughout the life what action should be repeated and which one should be avoided. This form of conditioning (operant conditioning) was first demonstrated by B.F. Skinner (1904–90) who an operant conditioning machine as a graduate student at Harvard University.[44] In an operant conditioning machine volunteers learn to press a key to receive reward. HM[43] and other individuals with lesions in the MTL are not only able to learn which key delivers reward, but also[45] learn consequences of their actions.

HM had intact motor learning, but he underperformed in tasks that required him to learn the sequence of motor movements by practicing repeatedly. HM did learn the sequence but he needed more practice than comparable control volunteers.[23] Additionally, HM had no linguistic impairment.[22] As mentioned above, language comprehension and linguistic expression involves nonconscious processing. These findings suggest that nonconscious memory and learning abilities are mostly retained after bilateral damage to the MTL.

HM's memory was impaired only for tasks that require conscious recollection of postsurgical events. He could not retain those memories for more than a few minutes. However, he performed well on tasks that did not require conscious recall of new memories. He had no difficulty retaining new nonconscious memories and retained the motor skills he learned in a mirror-tracing task for about a year.[46] These observations indicate that memory is not a unitary system and different forms of memories are processed in different brain areas. The MTL that was damaged in HM processes only certain types of conscious memory. Most nonconscious memories are not affected by MTL lesions.

Neural networks that process nonconscious memory

The observation that HM retained some forms of memory led investigators to study other amnesiac individuals. Those studies confirmed that conscious and nonconscious memories are processed in separate neural networks. It was a revolutionary idea and needed further study to explore networks that support two kinds of memory. Those studies were conducted using neuroimaging methods which are used to detect activity and localize activated brain areas during performance of a task. The initial neuroimaging study of implicit and explicit memory was conducted in the laboratory of neurologist Marcus Raichle at Washington University. His laboratory reported activation of the MTL during explicit memory processing and activation reduction in the occipital cortex in an implicit memory task.[47]

Around the same time, at the University of Oregon, we studied timeline of brain activation during processing of explicit and implicit memory using the event-related potential (ERP) technique. In this technique, the electrical activity of the brain, called electroencephalogram or EEG is recorded and using a computer algorithm, changes in EEG signals during performance of a task is detected. By detecting those changes, ERPs provide a timeline of the brain activation and an estimate of the brain areas involved in the processing. In these experiments we recorded the brain activity associated with encoding and recall under implicit and explicit memory conditions[38,48] using a modified version of the WSC task described earlier.[38,48]

For this experiment, a list of 450 words was prepared. Each word had a unique word stem (first three letters) and each stem could be completed using at least six words that are in common use in American English. Words were randomly divided into two lists (list A and B), each having 225 items. After EEG electrodes were placed on the scalp, volunteers were shown words of either list A or B. A few minutes later, word stems derived from words included in both A and B lists were presented. Volunteers were asked to make a word beginning with each stem. To elicit explicit memory, they were instructed to remember studied words and complete stems using a studied word. This made both encoding and retrieval a conscious or explicit process. To study implicit memory, volunteers were asked at the study stage to indicate whether they liked or disliked a presented word and were required to complete word stems using the first word that came to mind beginning with the stem. In both explicit and implicit conditions, volunteers were asked to respond as quickly and as accurately as possible. The number of stems completed using a studied word was computed and response time measured for each trial. As mentioned above, volunteers studied only one list, either A or B, but were asked to complete stems derived from words included in both lists because stems of nonstudied words were used as control. Thus, if list A was used as the study list, stems of list B words were used as control to determine the random chance of getting a correct response. Since list B was never shown, successful retrieval was not possible while completing stems derived from words of this list. It was possible only for completing stems of words derived from list A. Activation elicited during completion of stems from list B words was used as control both in the explicit and implicit conditions. Assignment of lists A and B as study and control was counterbalanced across volunteers. Thus, half of the volunteers studied list A and the other half studied list B.

In this experiment the patterns of brain activity observed during processing of conscious (explicit) and nonconscious (implicit) memory were different. When volunteers consciously recalled studied words, increased activity was observed in electrodes placed over the frontal cortex and in electrodes that receive signals from the MTL. Fig. 2.12 shows the location of those electrodes. To localize brain areas where changes in electrical activity originated, we used a dipole localization algorithm (brain electrical source analysis). We found increased activity in the left frontal cortex when a volunteer's effort to recall a word was unsuccessful. That means even when retrieval was not successful there was increased activity in the left frontal area. On the other hand, a successful retrieval activated both the right and left frontal cortex (Fig. 2.13). In addition, during conscious retrieval we observed

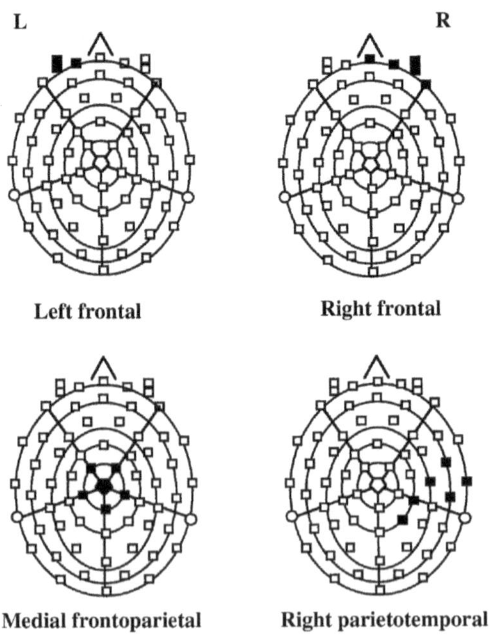

FIGURE 2.12 Placement of electrodes in the ERP experiment discussed. Image shows the location of the activated electrodes (*shaded*) under the different experimental conditions described in the text. *Reprinted from Badgaiyan RD, Posner MI. Time course of cortical activations in implicit and explicit recall.* J Neurosci *1997;**17(12)**:4904–13.*

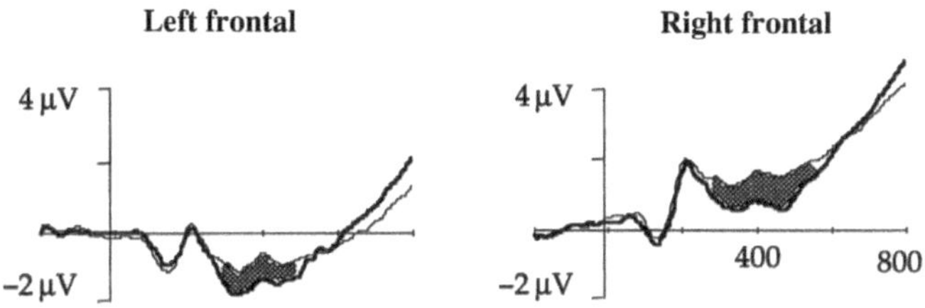

FIGURE 2.13 ERP showing increased activation in the left and right frontal cortex between 200 and 600 ms after stimulus presentation during explicit retrieval (thin lines), as compared to that observed during implicit retrieval (thick lines). *Reprinted from Badgaiyan RD, Posner MI. Time course of cortical activations in implicit and explicit recall.* J Neurosci *1997;**17**(12):4904–13.*

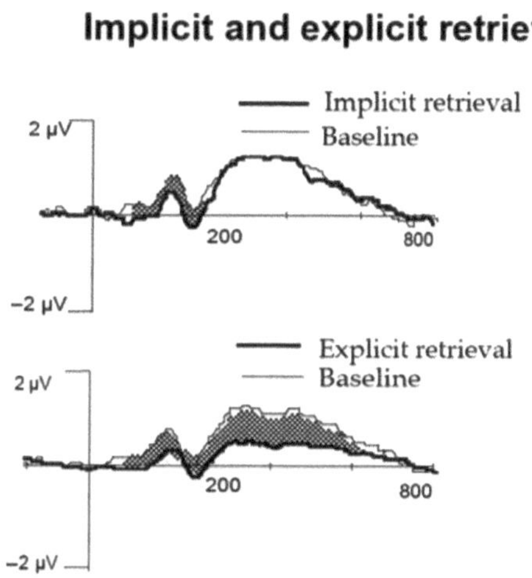

FIGURE 2.14 During nonconscious (implicit) retrieval reduced activation was observed in the posterior cortex (OPT junction area including area V3A), between 64 and 200 ms of stimulus presentation. The reduction was observed between 64 and 600 ms during conscious (explicit) retrieval in the same area. *Reprinted from Badgaiyan RD, Posner MI. Time course of cortical activations in implicit and explicit recall.* J Neurosci *1997;**17**(12):4904–13.*

reduced activation in the right posterior cortical electrodes (Fig. 2.12) located at the junction of the occipital, parietal, and temporal lobes, called the occipito-parieto-temporal (OPT) junction. This reduction was observed when the stems of studied words were presented, and not during presentation of the stems of nonstudied words. Reduced activation in the posterior cortex began very early, only 64 ms after stimulus presentation, and it ended around 600 ms (Fig. 2.14 bottom). Increased activation in the MTL began at around 150 ms and ended at 250 ms. The prefrontal activation began around 200 ms and ended at 600 ms (Fig. 2.13).

During implicit or nonconscious retrieval, there was reduced activation in the same posterior cortical area (OPT junction) where reduction was observed in the explicit condition. The duration of this reduction however, was shorter. It began at 64 ms but ended at 200 ms, as opposed to 600 ms in the explicit condition (Fig. 2.14). The frontal and MTL areas were not activated in implicit retrieval. These observations suggest that even though implicit and explicit memory are processed by separate neural networks, they share the processing area at OPT junction. This area served as the initial processing step during both conscious and nonconscious retrieval. In the implicit condition, it is the only area where a change in activity was observed, but in the explicit condition additional activity is observed in the frontal cortex and MTL. Since nonconscious retrieval needs only initial processing at the OPT junction, it is quicker and completes in 200 ms but explicit retrieval takes about 600 ms to complete. That is why nonconscious processing is quicker.

Because the results of these experiments were interesting, we decided to further investigate. The ERP technique used in these experiments does not accurately localize brain areas because activity recorded on the scalp surface sometimes yields misleading information about the location of activated area. Since these activities represent the sum of activities of the superficial and deep brain structures, it is difficult to isolate the area from where the activity originated. Additionally, surface activities are affected by synchronization of activities. Thus, if there is synchronized activity of the deep and superficial structures, surface potentials will have higher amplitude, falsely indicating enhanced activity. On the contrary, desynchronization attenuates or even masks the activity recorded on the scalp surface. It was therefore necessary to use another neuroimaging technique to localize the brain areas involved in processing of implicit and explicit memory.

We therefore used functional magnetic resonance imaging (fMRI) to validate the findings of the ERP experiment. The fMRI was the most reliable technique available at the time and is based on a method developed by Seiji Ogawa at Bell Labs in 1990.[49] This method, called blood oxygenation level dependent (BOLD) contrast, exploits the minute difference in magnetic properties of oxygenated and deoxygenated blood to detect changes in blood flow in different brain areas during a task performance.

In this experiment, after volunteers were shown a series of pictures for 3 seconds each, they were forced to nonconsciously retrieve them by briefly showing the same pictures again subliminally. Analysis of the data revealed reduced activation in the OPT junction, in the same location where reduction was observed during nonconscious retrieval in the ERP experiment discussed above. The fMRI experiment localized the area of reduced activation to the Brodmann area (BA) 19, with the most significant reduction located in a small area within BA 19, called area V3A (Figs. 2.15 and 2.20).[50] After other investigators found similar activation reduction in this area, it is generally accepted to be associated with nonconscious retrieval.

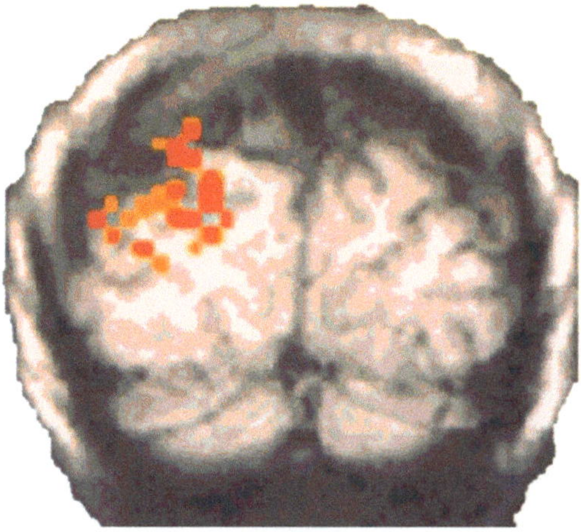

FIGURE 2.15 fMRI image of the brain showing BA 19 where reduced activation (marked *red* and *yellow*) was observed in a picture priming task. *Adapted from Badgaiyan RD. Neuroanatomical organization of perceptual memory: an fMRI study of picture priming.* Hum Brain Mapp *2000;**10**(4):197–203.*

Classically, area V3A is a part of the visual cortex and it is thus interesting that it would be involved in memory processing. Could the activity observed be associated with visual processing of stimuli rather than memory? We began exploring this question in Daniel Schacter's laboratory at Harvard University by designing a different kind of WSC task. In this task all stimuli were auditory, instead of visual. In the auditory version of the stem completion task volunteers listened to a list of words and decided if they liked/disliked them. They were then aurally presented the first syllable of studied and nonstudied words and asked to complete stems using a word that began with the presented syllable (auditory word stem). We made sure that each word in the list had a different first syllable and each syllable made at least six legal words. To our surprise we found reduced activation in the same brain area where reduction was observed in the visual WSC task.[37] In addition, there was also decreased activation in the medial frontal cortex (Fig. 2.16). The reduced activation in area V3A in this experiment confirmed that the reduction is associated with nonconscious retrieval and not with visual information processing.

The above experiment, while answering an important question concerning the role of area V3A, raised another question: what would happen if the words were studied aurally and volunteers retrieve them using visual word stems, or vice versa? We decided to examine this question by conducting two cross-modal priming experiments: auditory to visual and visual to auditory. In both experiments we found reduced activation in area V3A and increased activation in the prefrontal cortex (Fig. 2.17).[37,51]

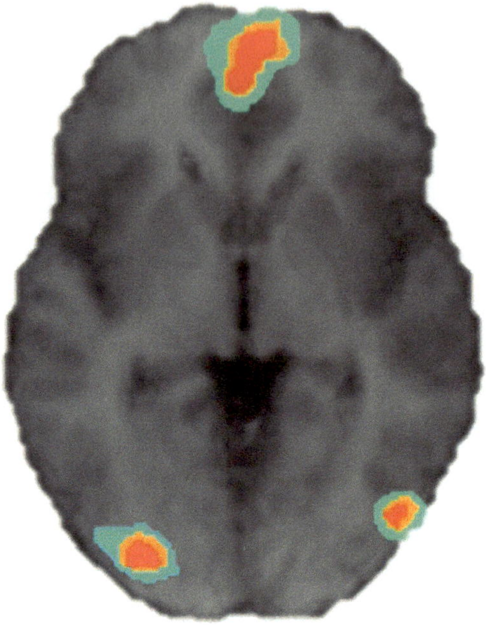

FIGURE 2.16 Reduced activation in area V3A (bottom) and in the medial prefrontal cortex (top) during auditory priming. *Adapted from Badgaiyan RD, Schacter DL, Alpert NM. Auditory priming within and across modalities: evidence from positron emission tomography.* J Cogn Neurosci *1999;**11**(4):337–48.*

The increased activation in the prefrontal cortex in the cross-modal priming experiments is interesting because this activation is associated with conscious retrieval and not with nonconscious retrieval.[38,50] To understand the significance of this activation we conducted two additional experiments.

In the first experiment we studied unimodal auditory priming but changed the voice from male to female between study and stem completion.[52] We again observed increased activation in the prefrontal cortex in addition to reduced activation in area V3A (Fig. 2.18). In the second experiment volunteers completed an auditory to visual cross-modal priming task in which they completed visually presented word stems after studying words aurally under full attention or divided attention conditions.[52] We decided to conduct this experiment because when attention is divided, conscious memory is impaired but not the nonconscious. This experiment therefore could help us understand whether the prefrontal activation observed in the previous experiment was due to involvement of conscious memory mechanisms (explicit contamination). In this experiment, at the study stage, volunteers were required to make only liking/disliking decisions on each word in full attention condition but in the divided attention condition, in addition to making liking/disliking decisions, they were also required to track a series of numbers that appeared

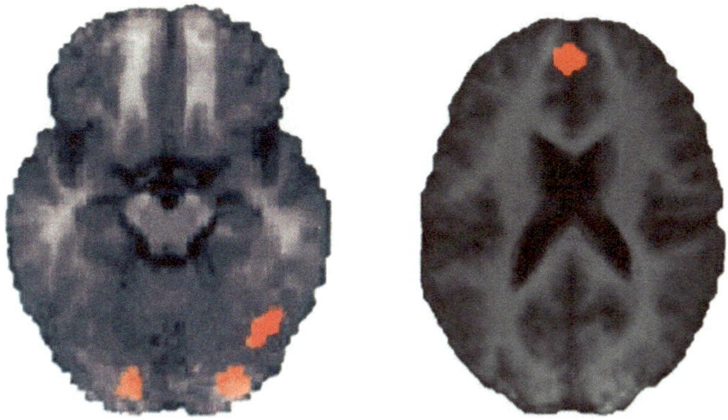

FIGURE 2.17 Reduced activation in area V3A (left) and increased activation in prefrontal area (right) during auditory-to-visual cross-modal priming. *Adapted from Schacter DL, Badgaiyan RD, Alpert NM. Visual word stem completion priming within and across modalities: a pet study.* Neuroreport *1999;**10**(10), 2061–2065.*

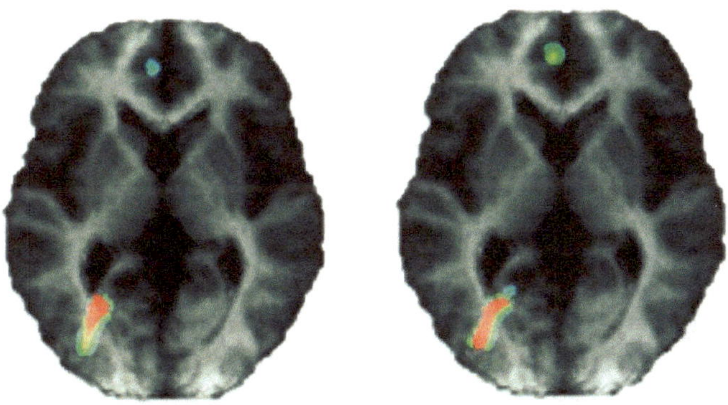

FIGURE 2.18 Reduced activation in area V3A in priming experiments in which either the same or different voices were used for encoding and retrieval. *Adapted from Badgaiyan RD, Schacter DL, Alpert NM. Priming within and across modalities: exploring the nature of rCBF increases and decreases.* Neuroimage *2001;**13**(2):272–82.*

on the monitor and to press a key when a number was repeated. We observed increased prefrontal activation in the full but not in the divided attention condition (Fig. 2.19). Since conscious or explicit memory was suppressed in the divided attention condition, the prefrontal activation disappeared. It indicated that the activation was associated with the explicit memory mechanism and its appearance in priming experiments suggests explicit memory contamination in the experiment.

The source of explicit memory contamination in cross-modal priming experiments could be the conceptual processing needed to complete the task.

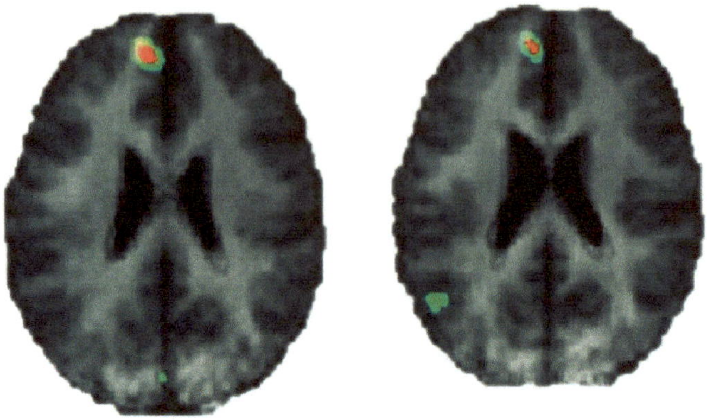

FIGURE 2.19 Activation in the prefrontal cortex was observed in priming under the full attention condition but not in the divided attention condition. The figures show two contrasts: Full Attention > Baseline (left) and Full Attention > Divided Attention (Right). *Adapted from Badgaiyan RD, Schacter DL, Alpert NM. Priming within and across modalities: exploring the nature of rCBF increases and decreases.* Neuroimage *2001;**13**(2):272–82.*

Because of the lack of a common perceptual feature between study items and word stems, stimuli are conceptually encoded in the cross modal condition, so that it can be retrieved using a cue presented in a different modality. Thus, for the brain to process information only on the basis of perceptual identification, stimuli presented in the study and the cue must be in the same modality. That is why a change in voice from male to female (described above) was not processed as perceptual but as conceptual priming. Becasue of the involvement of both, explicit and implicit mechanisms, the performance improves in the cross modal conditions. Thus, the rate of recall was higher in the full attention as compared to the divided attention condition.

Thus, the stem completion task can elicit either perceptual or conceptual priming depending on the perceptual similarity between the studied items and stems. Unimodal priming is mostly perceptual and cross-modal priming is almost always conceptual, involving components of the explicit mechanisms.

Area V3A and the gating mechanism

These experiments confirmed that the reduced activation in area V3A is associated with nonconscious memory processing. This area is in close proximity to the visual association areas that process visual stimuli. Visual association areas and are labeled as V2, V3, V4, V5 (Fig. 2.20). V1 is the primary visual cortex where stimuli from the retina converge for initial processing. From here signals are relayed to areas V2, V3, V4, and V5 for further processing. Not much is known about visual processing in area V3A partly because it was given a separate identity only in 1978 when David Van

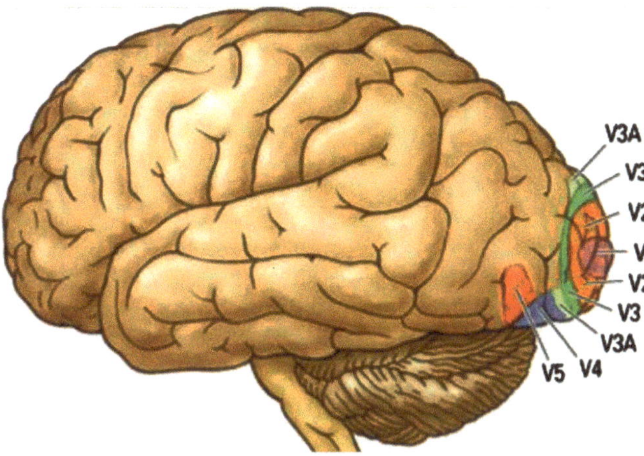

FIGURE 2.20 Location of the visual association areas V1, V2, V3, V3A, V4, and V5 in the human brain. *Adapted from the brain.mcgill.ca.*

Essen and S.M. Zeki at the University College, London noted that this area has distinct connections and morphology in monkeys.[53] Even though it is located between areas V3 and V4, it has little structural or functional similarity with either of these areas. Its afferent and efferent connections are entirely different from those of the other visual association areas that receive heavy input from area V1. Area V3A receives only sparse input from V1 and gets most of its input from the areas that process sensory information. That is why even though anatomically it is a part of the visual cortex, functionally it is a distinct entity.[54] Because of its rich connections with other brain areas, 30% of neurons of area V3A still respond to the light after the primary visual cortex (V1) is inactivated, while it completely shuts down activities in areas V2 and V3.[55] Thus, area V3A receives visual input from sources other than area V1. These sources include the superior colliculus, located in the midbrain and lateral geniculate nucleus of the thalamus. These inputs are responsible for blindsight in which individuals with a damaged primary visual cortex perceive visual stimuli without being consciously aware of it. This phenomenon is discussed in Chapter 3, Blindsight.

Area V3A is an important link between the visual cortex and rest of the brain. It helps to integrate visual inputs with inputs from other sensory modalities.[56] This integration helps the brain plan motor movements. Activity in area V3A is therefore observed before actual motor movement takes place. It is involved in planning of movements.[57] Because area V3A receives input from almost all sensory modalities and sends efferent connections to many cortical and subcortical areas, it is uniquely positioned to process nonconscious memory across sensory modalities.

Area V3A is involved in retrieval of relevant information during recall. Based on the data acquired in our experiments, it appears that this area has a gating mechanism. It allows retrieval of a limited amount of the most relevant information from previously encoded memory on an "as needed" basis. By releasing a limited amount of information, it helps the brain make quick decisions. We discovered this role of area V3A initially by analyzing data acquired in a series of ERP experiments described earlier.[38,48,58] In these experiments, after volunteers had studied a list of words, they were asked to recall under conscious and nonconscious conditions. We observed reduced activation in the posterior cortical areas (later identified as area V3A) during both conscious and nonconscious retrieval conditions. The reduced activation essentially had two phases—an early phase between 64 and 200 ms and a late phase beginning at 200 ms and ending at 600 ms. The early phase was associated with nonconscious retrieval because that was the only change we observed during implicit retrieval (Fig. 2.14). In this condition, volunteers were not consciously aware of retrieval, resulting in study participants denying studying words they generated using word stems. Under the conscious condition volunteers had to be fully aware of retrieval. They knew that they had retrieved a studied word. Therefore, cognitively, conscious awareness of retrieval was the only difference between conscious and nonconscious conditions. The additional brain activity observed in the conscious condition should therefore be responsible for bringing nonconsciously retrieved information to conscious awareness. These activities were the late phase of reduced activation in the posterior cortex including area V3A (between 200 and 600 ms of stimulus presentation) and increased activation in the MTL (between 150 and 200 ms) and frontal cortex (between 200 and 600 ms). One or both of these activities should therefore be associated with conscious awareness.[58]

The timeline of the late phase of reduced activation at the posterior cortex and that of the increased activation in the frontal cortex provides a clue to a possible neural mechanism associated with conscious awareness. A comparison of activities recorded during conscious and nonconscious conditions depicted in Fig. 2.21 shows a perfect overlap of the timeline of two activities between 200 and 600 ms after stimulus presentation. The overlap indicates a two-way exchange of information between these areas. Thus, in memory tasks after relevant information is retrieved at area V3A nonconsciously, a reentrant loop between area V3A (posterior cortex) and the frontal cortex is activated when conscious awareness of retrieval is needed (in explicit memory tasks). This process takes about 400 ms (between 200 and 600 ms), which is consistent with earlier observations suggesting that a stimulus needs to be processed for about half a second before it can be consciously perceived.[59] It is also consistent with a theoretical model proposed by the codiscoverer of the structure of DNA and Nobel laureate, Francis Crick, and neuroscientist Christof Koch.[60] The model suggests that a reentrant loop has to be established between the anterior and posterior cortical areas to make a stimulus available to conscious awareness. Recent findings of a direct

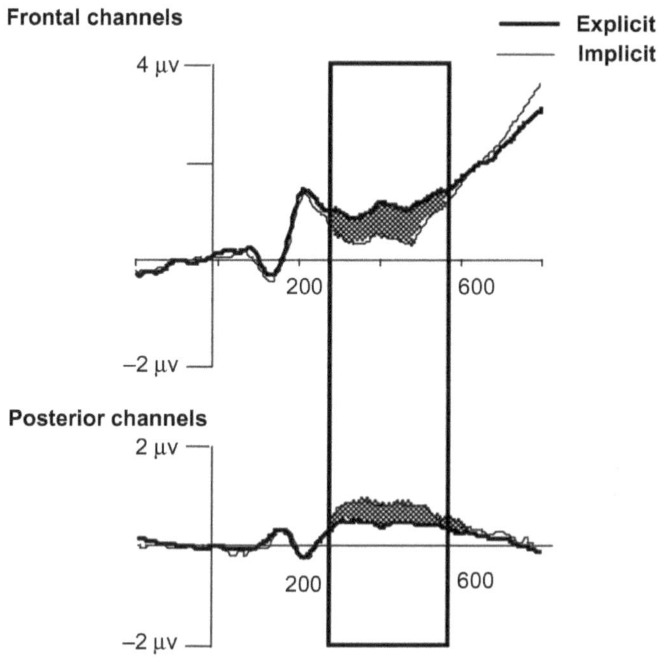

FIGURE 2.21 Contrast of ERP obtained under conscious (explicit) and nonconscious (implicit) conditions suggest perfect overlap of potential changes between posterior (including area V3A) and frontal channels, indicating activation of a reentrant loop between the two areas. *Reprinted from Badgaiyan RD. Conscious awareness of retrieval: an exploration of the cortical connectivity.* Int J Psychophysiol *2005;**55**(2):257–62.*

pathway between area V3A and the prefrontal cortex in monkeys further support the existence of the loop between area V3A and the frontal cortex.[61] It is also possible that the dorsal and ventral pathways that connect the visual cortex (including area V3A) and the frontal cortex carry fibers of the V3A frontal loop. These pathways are discussed in Chapter 3, Blindsight.

Because the V3A frontal loop makes us consciously aware of retrieved information, a lesion in the frontal cortex impairs awareness of perceived stimuli in a number of ways including increased temporal threshold for perception.[62] That is why a stimulus presented for a short period of time and recognizable to a healthy person becomes unrecognizable after a lesion in the frontal cortex.[63] Additionally, stimulation of the frontal cortex in the frontal eye field area by transcranial magnetic stimulation decreases the temporal threshold for visual perception and makes normally unrecognizable subliminal stimuli recognizable. Predictably, suppression of the dorsolateral prefrontal cortex increases the temporal threshold with the opposite effect.[64]

Primarily because of additional processing in the V3A frontal loop, conscious processing is slower than nonconscious processing. Additionally, conscious processes also involve activation of the attentional system, which probably contributes to the delay. This system is probably activated while information is processed in the V3A frontal loop.[65] As discussed in Chapter 3, Blindsight, and Chapter 4, Hemineglect, this loop passes through the areas (parietal and frontal cortex) involved in attentional processing.

Since the attentional system takes up to 500 ms to process a stimulus,[66] it may be responsible for the relatively long processing time in the V3A frontal loop. Additionally, because the attention system cannot process multiple stimuli simultaneously, if the brain must consciously process two stimuli and if a second stimulus is presented within "refractory period" of the first, it remains unprocessed. The refractory period is about 500 ms,[66] but it could be shorter under certain conditions. A review of the literature suggests that if two stimuli are presented within 300 ms of each other (stimulus-onset asynchrony, SOA), the attentional system is not activated. This time period therefore is within the refractory period. Depending on the task, attentional networks begin to activate when the SOA exceeds 300 ms.[67] Our observation of 400 ms of processing time in the V3A frontal loop therefore could be the time attentional networks need to process a stimulus.[67] However, this time period may be affected by many factors including type, intensity and contrast of stimuli, interfering or facilitating factors, and mental status of the volunteer.

It is not clear how attentional networks are activated in memory tasks, but it is possible that the V3A frontal loop activates the system either in the frontal or parietal cortex or both.[68] The V3A frontal loop passes through both of these structures. Damage to this loop may be the reason why people with a lesion in the parietal cortex deny awareness of stimuli they consciously perceive. This condition, called hemineglect, is discussed in Chapter 4, Hemineglect.

If area V3A is responsible for retrieval of retained information, as we have proposed, what happens if this area is damaged? After surgical resection of the cortex that would damage the rat equivalent of area V3A, the animal's memory is disrupted.[69] This finding is consistent with the proposed role of area V3A in recall. A human volunteer with damage limited to area V3A has not been studied because it is extremely rare to have a circumscribed lesion. However, a person with bilateral occipital damage (including area V3A) was studied by Suzanne Corkin (Fig. 2.8).[70] After the damage this individual had difficulty generating exemplars from a category (e.g., generating words like dog or cat for the animal category) but had "better than normal" performance in both unimodal visual priming and in auditory-to-visual cross-modal priming tasks.[70] Observation of a better-than-normal priming following a lesion involving area V3A is a significant finding in context of the proposed role of this area. If area V3A controls release of retained information, a lesion would disrupt this controlled release, and would result in flooding of information (both relevant and irrelevant) that would impair conscious but improve nonconscious memory. Conscious memory would be impaired because the brain must sort through a larger volume of information released to select intended item. Moreover, because conscious processes involve the attentional system, which has a refractory period, these processes can handle a limited number of information in a given time.[65] Because of this reason flooding of information would make conscious retrieval difficult, if not impossible. It would however not affect nonconscious processing because there is no limit to the number of options it can examine in a given time because the rate-limiting attention system is not involved in nonconscious processing.[65]

In fact, flooding of information could make nonconscious processes relatively more efficient because these processes are better at selecting the right option when the pool of options is large. This rather counterintuitive effect was experimentally demonstrated by a Dutch psychologist Albert Dijksterhuis and his colleagues.[71] In their experiment, volunteers were given four attributes (price, gas mileage, etc.) of four different cars. They were then asked to decide either immediately or after deliberating for four minutes which car is the best. As expected, responses were better when they had time to analyze attributes. However, when the experiment was repeated, and instead of four, volunteers were given 12 attributes of each car, responses were better when decisions were made immediately. The immediate recall condition was primarily a nonconscious condition because volunteers had no time to consciously analyze all 12 attributes and make a decision. Decisions were was based on a "gut feeling." When volunteers were given four minutes to think, conscious mechanisms were involved. The results of this experiment suggest that conscious decisions are better when the number of options is limited, but when many options are involved, nonconscious decisions are better. These results are convincing but need to be interpreted cautiously because investigators have failed to replicate many of Dijksterhuis' data.[72]

Irrespective of the controversy concerning this experiment, it is clear that explicit mechanisms cannot efficiently process a large pool of information because it involves the attentional system,[65] which has a long refractory period.[73] That is why flooding of information following damage to area V3A is likely to impair the conscious but not the nonconscious memory.

This newly discovered function of area V3A discussed above and in our earlier publications[37,50,52,67] helps us understand the neuropathology of many psychiatric conditions. Since conscious cognition, behavior, and action are dependent on nonconsciously retrieved information, inappropriate nonconscious retrieval would affect these functions significantly, leading to psychiatric symptoms. As discussed earlier, even a simple task of making conversation depends on nonconsciously retrieved words, grammar, syntax, intonation, and other linguistic attributes. Similarly, our response to a situation depends on nonconsciously retrieved memory of past experiences, knowledge of similar situations, and culturally acceptable norms. All of these and many more pieces of information contribute to our actions and behaviors even though we do not consciously know what information was retrieved and used nonconsciously while dealing with a situation.

Nonconsciously retrieved information also influences our behavior by optimizing perception and cognition. We know sensory stimuli received in our sense organs are modified at several stages before we perceive and understand them. That is why perception is not an exact replica of the signals sensory organs receive. These signals are modified based on experience, cultural norms, beliefs, previous learning, and a host of other factors. That is why we perceive a rail track as parallel lines even though its image at retina appears to converge at a distance. Similarly, the red bar in Fig. 2.22 appears bigger at the far side despite having the exact same length. Because of this modification, we perceive smiles as faces, even though they lack many facial features. The brain extrapolates the perceived image and puts it in a context we can understand. Because of this extrapolation, a picture of the head of a dog is perceived as a whole dog. The head is extrapolated to a whole animal nonconsciously because experience tells us that the head is always associated with the body. Due to this extrapolation every day we recognize objects by looking at only a part of it (e.g., a phone tucked under a pile of paper, a cup hidden behind a pot, etc.).

While nonconscious modification has obvious advantages, it also leads to distorted perception and memory because we perceive something different

FIGURE 2.22 The red bar at the top looks longer than the one at the bottom, even though they have the exact same length.

from what our sensory system has received. This distortion was demonstrated by a Cambridge University psychologist Sir Frederick Charles Bartlett (1886–1969). In his book *Remembering: A Study in Experimental and Social Psychology* published in 1932,[74] Bartlett described an experiment in which volunteers read an Indian folklore called "The War of the Ghosts" and recalled it repeatedly. He claimed that the story got distorted with each retrieval and after several repetitions there were many distortions. Interestingly, this conclusion was at odds with the results of another experiment published in 1913[75] by another British psychologist Philip Boswood Ballard (1865–1950). Ballard had found that after repetition, schoolchildren were able to recall a poem more accurately. Even though Ballard was a well-respected psychologist and his books on mental tests were immensely popular, his experiments on repetition did not get the attention they deserved. In recent years, researchers have tried to replicate the two experiments to understand which of the contradicting conclusions is correct. Most experiments support Ballard's findings. When Mark Wheeler and Henry Roediger III of Rice University used the same text "The War of the Ghosts," they found that repetition actually improved accuracy of recall.[76] Bartlett's findings may not be replicable, but he introduced the concept of "reproductive memory," which refers to the memory of an event as perceived by a person.

Sometimes reproductive memory is inaccurate despite accurate perception. This phenomenon is called false memory, which is a normal phenomenon and can be tested using one of several sets of lure words constructed by Roediger and his student Kathleen McDermott.[77] To test, read the following words rapidly only once: *sour, candy, sugar, bitter, good, taste, tooth, nice, honey, soda, chocolate, heart, cake, tart, pie*. Do not look back. Wait for a minute and try to recall if the word *sweet* was in the list. Most people would incorrectly say that it was. This is an example of false memory and it happens because stimuli are stored in the brain nonconsciously as "concepts." Since a concept is dependent on past experiences, biases, and personal beliefs, it distorts the original stimulus. In the above example, the list had many words that are semantically related to the target word *sweet*. Therefore, we believe that the word sweet was a part of the list. Obviously, this distortion happens without conscious knowledge. It is a nonconscious phenomenon.

Nonconscious memory and psychiatric symptoms

Nonconscious retrieval therefore affects our perception. If retrieval is impaired, an object in the visual field could be misidentified or a familiar object not recognized at all, leading to cognitive deficits. Extrapolation of a sensory stimulus as discussed above can also be done consciously. It would then require us to consciously match the stimulus with all objects we have seen or known, to be able to recognize it. Obviously, this process would fail because either it would not find an exact match (and exact view) or it would take incredibly long time to choose one item from millions of objects and views we might have ever seen or known about. This process would come up either with too many possible matches, or no match at all, leading to false perception and altered cognition, behavior, and action. Since these alterations are key features of psychiatric conditions, neurocognitive deficits in some of these conditions could possibly be traced to impaired nonconscious retrieval. It is an uninvestigated but important and promising area of research.

In a recent publication we proposed that the link between symptoms of schizophrenia and impaired functions of area V3A should be examined closely.[67] We argued that many clinical symptoms of schizophrenia could be produced if nonconscious retrieval is impaired by damaged area V3A. As discussed above, damage to this area would result in release of a large volume of information from memory, both relevant and irrelevant to a situation. This would lead to a flood of information, confusion, and disorganized thoughts. It would also leave some of the retrieved information unprocessed or partially processed, leading to source confusion or source misattribution.

Source attribution is an important aspect of memory. It gives context to an event and helps us remember not only the event but also whether it was an experienced or imagined event. Misattribution of a source would lead us to believe that an imagined or fictional event was an experienced event or vice versa. Source confusion and source misattribution are common in children because of their poorly developed frontal cortex, which is involved in processing source attribution. Because of source misattribution children often think a character in a story is a real character they have seen or met. Misattribution happens in adults too when memory encoding of an event is incomplete. For some strange and unclear reason, sometimes it also happens collectively in a group of people.

Misattributed memory has been responsible for a number of lawsuits in the United States. In these cases, a child or a group of children have "recalled" memories of sexual assault. The most infamous of those cases was the McMartin preschool trial. It was the longest and most expensive criminal trial in the US history. It started in 1983 when several preschool children accused their teachers of sodomizing and molesting them. It turned out that the accusations were a result of suggestive interviews conducted in a Los Angeles clinic. During the interviews children were asked to pretend abuse. On repeated suggestions, they misattributed the suggestions and believed those events actually happened. There was no physical evidence and no corroboration. It started with a complaint by a woman named Judy Johnson who "believed" her son was sodomized by a teacher. Later, it was discovered that she was a chronic alcoholic and suffered from paranoid schizophrenia. Her belief was based on paranoia and compromised mental health. The trial ended in 1990 with no conviction.

Source confusion was also responsible for a flood of stories of alien abduction reported by a number of people in the 1970s, mostly in New

Hampshire.[78] It was also responsible for "recovered" memories of many American women in the 1980s. They claimed to have recovered memories of being forced by a cult to participate in satanic rituals that included sexual assault. None of those stories has ever been verified. Interestingly, it began after publication of a popular book called *Michelle Remembers* by Michelle Smith and Lawrence Pazder, published in 1980 on a similar theme (Fig. 2.23).[79] Apparently, after reading the book, those women misattributed the source. They thought that experiences of the book's character were their own. If an imagined experience is misattributed as a real experience, one can have auditory or visual hallucination, which is a feature of many psychiatric and neuropsychiatric conditions.

As discussed above, damage to area V3A would enhance implicit and impair explicit memory. That is what happens in schizophrenia. In a largely ignored experiment, a Boston psychiatrist named Theo Manschreck observed that individuals with thought disorders (including schizophrenia) perform

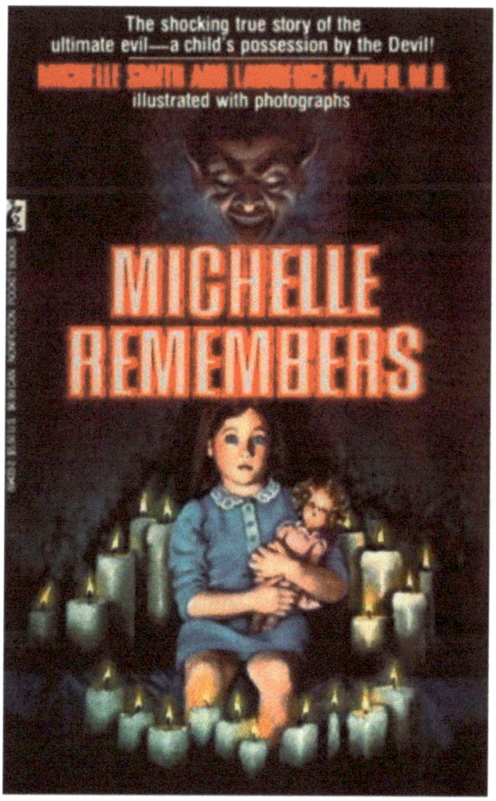

FIGURE 2.23 The book that prompted many women to "recover" memories of forceful participation in satanic rituals. *Image reproduced from Wikimedia Commons.*

better than healthy control volunteers in lexical decision taskdiscussed earlier.[80] In this task, they recognize the target word more quickly than healthy control volunteers if it was primed with a semantically related word (e.g., the word doctor primed with the word nurse). Since, this study included only 12 volunteers it did not get the attention it deserved. Later, two psychologists Richard Linocut and Robert Knight of New Zealand showed a similar effect in a larger cohort of people with schizophrenia.[81] These findings were consistent with a damaged area V3A in these patients. Deficits in visual processing and attention shifting observed in schizophrenia also indicate damage to area V3A.[82,83] A functionally damaged area V3A could also account for other symptoms of schizophrenia. These symptoms include hallucination and disorganized thoughts as discussed above. Delusions or false beliefs could also be due to incomplete processing of a large amount of information released by a damaged V3A.[67]

Depending on the nature of damage, it is possible for a damaged V3A to excessively limit or completely block release of information when needed. If that happens, the person will not have input for cognitive processing and for making decisions. This could then lead to the so-called negative symptoms observed in a subset of patients with schizophrenia and other neuropsychiatric conditions. These patients lack motivation, emotional responsiveness, and motor movements. All of these symptoms could be theoretically produced if data needed to process emotion, social interaction, or actions are not retrieved nonconsciously.

If area V3A is damaged in schizophrenia, do we have an evidence to show? Yes, but because of the limitations of our neuroimaging methods, a small damaged area is difficult to detect, particularly if the damage is functional. Indirect evidence that suggests the damage includes finding of a weak negative potential 150 ms after stimulus presentation (N150 wave) in ERP recordings of schizophrenia. This finding is important because N150 waves originate in areas V3 and V3A.[84] Further, it was suggested that abnormal activity (oscillations) in area V3A triggers a sequence of events that leads to clinical symptoms of schizophrenia.[85] Studies on visual processing in schizophrenia have also reported functional damage to area V3A.[86] Additionally, in schizophrenia, structural[87] and functional[88] abnormalities have been reported in the medial and superior temporal gyri, which are in close proximity to area V3A. These abnormalities could interrupt signals going in or out of area V3A. It is also possible that area V3A is functionally damaged in these individuals at birth because it is highly sensitive to hypoxia during development. In addition to causing structural and functional damage, hypoxia is also known to make a person vulnerable to environment-induced changes in genetic signals called epigenetic changes. These changes are known to alter brain activity by turning specific genes "on" or "off," These alterations are believed to cause many psychiatric conditions including schizophrenia.[85] There is therefore a need to study the nature of damage to area V3A in schizophrenia and other psychiatric conditions. It could well be the primary target for therapeutic intervention in these conditions.

As mentioned above, individuals with schizophrenia process implicit memories more efficiently than healthy individuals. However, it is not true for all tasks of implicit memory. In many of these tasks their performance is suboptimal.[67,89] It is therefore important to understand in what kind of tasks they perform better and why. Our analysis revealed that they perform better than healthy volunteers in experiments that have relatively short intervals between two consecutive stimuli, called stimulus onset asynchrony or SOA.[67] Individuals with schizophrenia perform better than healthy controls in experiments that use a short SOA of less than 300 ms. Their performance begins to deteriorate as SOA increases.[67]

So why does SOA affect performance? It could be due to involvement of the attentional system. As mentioned earlier, the attentional system requires time to process a stimulus. Michael Posner and Charles Snyder estimated this time to be approximately 500 ms.[66] Based on our data, it is about 400 ms.[65] This means that in order for the attentional system to attend to a stimulus, a second stimulus must be presented after the first stimulus has been completely processed. As discussed earlier, the attentional system needs to be activated to make a task conscious or explicit.[65] Therefore, experiments with longer SOAs activate the attentional system and make the task explicit. People with schizophrenia perform better than healthy volunteers if the task does not require activation of the attentional system. This makes sense because attentional processes are impaired in schizophrenia.

Data acquired in our laboratory and elsewhere suggest that there are at least three components of memory retrieval: Nonconscious retrieval, which is retrieval without attention and conscious awareness; nonconscious retrieval with activation of the attentional system but without conscious awareness; and conscious retrieval in which the attentional system is activated and there is conscious awareness of retrieval. Thus, full conscious retrieval involves at least three steps: nonconscious retrieval, activation of the attentional system, and activation of the neural networks that bring the retrieval to conscious awareness. These steps are usually processed sequentially. A nonconscious memory task involves only retrieval without attention and awareness, or it can activate retrieval with attention but without awareness.

Designing an implicit task therefore is tricky because sometimes it is a challenge to avoid explicit contamination. Strategies to achieve that include the use of a long study list. Thus, making volunteers study 100 items before asking them to retrieve is a better strategy than giving them only 10 items to study. Since it is relatively easy to recall the first and last items of a list because of primacy and recency effects, a good design uses "filler" items at the beginning and end of the list. These items are not used for retrieval. Nonconscious processing can also be assured using techniques like the process dissociation paradigm developed by psychologist Larry Jacoby.[90] In this paradigm, after volunteers are shown a list of study items they are tested under "inclusion" and "exclusion" conditions. In the inclusion condition they must recall study list items while in the exclusion condition they must avoid items in the study list. The use of a studied item in the exclusion condition confirms nonconscious retrieval because conscious processing would have prevented its use. The use of a studied item in the inclusion condition could either be due to conscious or nonconscious retrieval. By comparing responses in the two conditions, the extent of explicit contamination can be estimated. Another strategy to make a task nonconscious is to keep SOAs as short as possible—preferably less than 300 ms. It is the most effective strategy because it ensures that the brain networks that control attention and conscious awareness do not have adequate time to process the stimulus.

The data discussed in this chapter suggest that we cannot perform some of our basic activities of daily life without nonconscious memory. More significantly, nonconscious processes play an important role in the phenomenology of psychiatric and neuropsychiatric conditions. This aspect, however, has not yet been fully appreciated and investigated.

Bibliography

1. Myers GC. A study of incidental memory. *Arch Psychol* 1913;**21**(4) (Suppl. 26):21–108.
2. Cattell JM. Measurements of the accuracy of recollection. *Science* 1895;**2**(49):761–6.
3. Binet A. Mnemonic virtuosity: a study of chess players. In: *Genetic psychology monographs* 1966;**74**:127–162.
4. Shellow SR. *Individual differences in incidental memory*. Columbia University; 1923.
5. Jenkins JG. Instruction as a factor in 'incidental' learning. *Am J Psychol* 1933;**45**(3):471–7.
6. Reber AS. Implicit learning of artificial grammar. *J Verb Learn Verb Behav* 1967;**6**:855–63.
7. Tilley AJ. Sleep learning during stage 2 and REM sleep. *Biol Psychiatry* 1979;**9**(3):155–61.
8. Cooper LM, Hoskovec J. Hypnotic suggestions for learning during stage I REM sleep. *Am J Clin Hypnosis* 1972;**15**(2):102–11.
9. Wood JM, Bootzin RR, Kihlstrom JF. Implicit and explicit memory for verbal information presented during sleep. *Psychol Sci* 1992;**3**(4):236–40.
10. Kihlstrom JF, Schacter DL, Cork L, Hurt CA, Behr SC. Implicit and explicit memory following surgical anesthesia. *Psychol Sci* 1990;**1**(5):303–6.
11. Merikle PM, Daneman M. Memory for unconsciously perceived events: evidence from anesthetized patients. *Conscious Cogn* 1996;**5**(4):525–41.
12. Cork RC, Kihlstrom JF, Schacter DL. Absence of explicit or implicit memory in patients anesthetized with sufentanil/nitrous oxide. *Anesthesiology* 1992;**76**(6):892–8.
13. Smith TL, Maye JP. Evaluation of word associations and the bispectral index monitor as indicators of implicit memory formation during general anaesthesia. In: Jordan C, Vaughan DJA, Newton DEF, editors. *Memory and* awareness in anaesthesia *IV*. World Scientific; 1998. p. 203–9.
14. Penfield W. Memory mechanisms. *AMA Arch Neurol Psychiatry* 1952;**67**(2):178–98.

15. Harris TA. *I'm OK, You're OK: a practical guide to transactional analysis.* Harper & Row; 1969.
16. Akimenko M. Vladimir Mikhailovich Bekhterev. *J Hist Neurosci* 2007;**16**(1–2):90–9.
17. Ogden JA, Corkin S. Memories of H. M. In: *Memory Mechanisms: A Tribute to G. V. Goddard* Eds. W.C. Abraham, M Corballis, & K.G. White, Lawrence Erlbaum Associates 1991, pp. 195–215.
18. Milner B, Penfield W. *The effect of hippocampal lesions on recent memory. Transactions of the American Neurological Association (80th Meeting).* 1955. p. 42–8.
19. Milner B. The memory defect in bilateral hippocampal lesions. *Psychiatr Res Rep Am Psychiatr Assoc* 1959;**11**:43–58.
20. Milner B. Memory disturbance after bilateral hippocampal lesions. In: Milner PM, Glickman S, editors. *Cognitive processes and the brain.* Van Nostrand; 1965. p. 97–111.
21. Corkin S. Tactually-guided maze learning in man: effects of unilateral cortical excisions and bilateral hippocampal lesions. *Neuropsychologia* 1965;**3**:339–51.
22. Milner B, Corkin S, Teuber HL. Further analysis of the hippocampal amnesic syndrome: 14-year follow-up study of H.M. *Neuropsychologia* 1968;**6**:215–34.
23. Corkin S. Acquisition of motor skill after bilateral medial temporal-lobe excision. *Neuropsychologia* 1968;**6**(6):255–65.
24. Scoville WB, Milner B. Loss of recent memory after bilateral hippocampal lesions. *J Neurosurg Psychiatry* 1957;**20**:11–21.
25. Sagar HJ, Cohen NJ, Corkin S, Growdon JH. Dissociations among processes in remote memory. *Ann N Y Acad Sci* 1985;**444**:533–5.
26. Corkin S. What's new with the amnesic patient H.M.? *Nat Rev Neurosci* 2002;**3**(2):153–60.
27. Storms LH. *Backward association in verbally mediated learning* (Ph.D. thesis), University of Minnesota; 1956.
28. Storms LH. Apparent backward association: a situational effect. *J Exp Psychol* 1958;**55**(4):390–5.
29. Cofer CN, Ford TJ. Verbal context and free association-time. *Am J Psychol* 1957;**70**(4):606–10.
30. Cramer P. Successful mediated priming via associative bonds. *Psychol Rec* 1964;**15**:235–8.
31. Meyer DE, Schvaneveldt RW. Facilitation in recognizing pairs of words: evidence of a dependence between retrieval operations. *J Exp Psychol* 1971;**90**(2):227–34.
32. Landauer TK, Freedman JL. Information retrieval from long term memory: category size and retrieval time. *J Verbal Learn Verbal Behav* 1968;**7**:291–5.
33. Scarborough DL, Cortese C, Scarborough HS. Frequency and repetition effects in lexical memory. *J Exp Psychol: Human Percept Perform* 1977;**3**:1–17.
34. Gabrieli JDE, Keane MM, Stanger BZ, Kjelgaard MM, et al. Dissociations among structural-perceptual, lexical-semantic, and event-fact memory systems in Alzheimer, amnesic, and normal subjects. *Cortex* 1994;**30**(1):75–103.
35. Warrington EK, Weiskrantz L. Amnesic syndrome: consolidation or retrieval? *Nature* 1970;**228**(5272):628–30.
36. Badgaiyan RD. Executive control, willed actions, and nonconscious processing. *Hum Brain Mapp* 2000;**9**(1):38–41.
37. Badgaiyan RD, Schacter DL, Alpert NM. Auditory priming within and across modalities: evidence from positron emission tomography. *J Cogn Neurosci* 1999;**11**(4):337–48.

38. Badgaiyan RD, Posner MI. Time course of cortical activations in implicit and explicit recall. *J Neurosci* 1997;**17**(12):4904–13.
39. Graf P, Schacter DL. Implicit and explicit memory for new associations in normal and amnesic subjects. *J Exp Psychol: Learn Mem Cogn* 1985;**11**(3):501–18.
40. Gabrieli JD, Milberg W, Keane MM, Corkin S. Intact priming of patterns despite impaired memory. *Neuropsychologia* 1990;**28**(5):417–27.
41. Gabrieli JD, Cohen NJ, Corkin S. The impaired learning of semantic knowledge following bilateral medial temporal-lobe resection. *Brain Cogn* 1988;**7**(2):157–77.
42. Woodruff-Pak DS. Eyeblink classical conditioning in H.M.: delay and trace paradigms. *Behav Neurosci* 1993;**107**(6):911–25.
43. Sidman M, Stoddard LT, Mohr JP. Some additional quantitative observaitons of immediate memory in a patient with bilateral hippocampal lesions. *Neuropsychologia* 1968;**6**:245–54.
44. Skinner BF. A discrimination without previous conditioning. *Proc Natl Acad Sci USA* 1934;**20**(9):532–6.
45. Channon S, Daum I, Gray JA. Operant conditioning after temporal lobe lesions in man: conditional and simple discrimination learning. *Cortex* 1993;**29**(2):315–24.
46. Gabrieli JD, Corkin S, Mickel SF, Growdon JH. Intact acquisition and long-term retention of mirror-tracing skill in alzheimer's disease and in global amnesia. *Behav Neurosci* 1993;**107**(6):899–910.
47. Buckner RL, Petersen SE, Ojemann JG, Miezin FM, Squire LR, Raichle ME. Functional anatomical studies of explicit and implicit memory retrieval tasks. *J Neurosci* 1995;**15**(1 Pt 1):12–29.
48. Badgaiyan RD, Posner MI. Priming reduces input activity in right posterior cortex during stem completion. *Neuroreport* 1996;**7**(18):2975–8.
49. Ogawa S, Lee TM, Kay AR, Tank DW. Brain magnetic resonance imaging with contrast dependent on blood oxygenation. *Proc Natl Acad Sci USA* 1990;**87**(24):9868–72.
50. Badgaiyan RD. Neuroanatomical organization of perceptual memory: an fMRI study of picture priming. *Hum Brain Mapp* 2000;**10**(4):197–203.
51. Schacter DL, Badgaiyan RD, Alpert NM. Visual word stem completion priming within and across modalities: a pet study. *Neuroreport* 1999;**10**(10):2061–5.
52. Badgaiyan RD, Schacter DL, Alpert NM. Priming within and across modalities: exploring the nature of rCBF increases and decreases. *Neuroimage* 2001;**13**(2):272–82.
53. Essen DC, Zeki SM. The topographic organization of rhesus monkey prestriate cortex. *J Physiol* 1978;**277**:193–226.
54. Tootell RB, Mendola JD, Hadjikhani NK, Ledden PJ, Liu AK, Reppas JB, et al. Functional analysis of V3A and related areas in human visual cortex. *J Neurosci* 1997;**17**(18):7060–78.
55. Girard P, Salin PA, Bullier J. Visual activity in areas V3A and V3 during reversible inactivation of area V1 in the macaque monkey. *J Neurophysiol* 1991;**66**(5):1493–503.
56. Ogawa A, Macaluso E. Audio-visual interactions for motion perception in depth modulate activity in visual area V3A. *Neuroimage* 2013;**71**:158–67.
57. Nakamura K, Colby CL. Visual, saccade-related, and cognitive activation of single neurons in monkey extrastriate area V3A. *J Neurophysiol* 2000;**84**(2):677–92.
58. Badgaiyan RD. Conscious awareness of retrieval: an exploration of the cortical connectivity. *Int J Psychophysiol* 2005;**55**(2):257–62.
59. Libet B. The neural time factor in conscious and unconscious events. *Ciba Found Symp* 1993;**174**:123–37 discussion 137–46.

60. Crick F, Koch C. Are we aware of neural activity in primary visual cortex? *Nature* 1995;**375**(6527):121–3.
61. Stanton GB, Bruce CJ, Goldberg ME. Topography of projections to posterior cortical areas from the macaque frontal eye fields. *J Compar Neurol* 1995;**353**(2):291–305.
62. Del Cul A, Dehaene S, Reyes P, Bravo E, Slachevsky A. Causal role of prefrontal cortex in the threshold for access to consciousness. *Brain* 2009;**132**(Pt 9):2531–40.
63. Wiener M, Coslett HB. Disruption of temporal processing in a subject with probable frontotemporal dementia. *Neuropsychologia* 2008;**46**(7):1927–39.
64. Grosbras MH, Paus T. Transcranial magnetic stimulation of the human frontal eye field facilitates visual awareness. *Eur J Neurosci* 2003;**18**(11):3121–6.
65. Badgaiyan RD. Nonconscious perception, conscious awareness and attention. *Conscious Cogn* 2012;**21**(1):584–6.
66. Posner MI, Snyder CRR. Attentional and cognitive control. In: Solso R, editor. *Cognitive psychology*. Lawrence Erlbaum Associates; 1975. p. 55–85.
67. Badgaiyan RD. Nonconscious processing and a novel target for schizophrenia research. *Open J Psychiatry* 2012;**2**(4A).
68. Posner M, Badgaiyan R. Attention and neural networks. In: Parks RW, Levine DS, editors. *Fundamentals of neural network modeling: neuropsychology and cognitive neuroscience*. The MIT Press; 1998. p. 61–76.
69. Kesner RP, Bolland BL, Dakis M. Memory for spatial locations, motor responses, and objects: triple dissociation among the hippocampus, caudate nucleus, and extrastriate visual cortex. *Exp Brain Res* 1993;**93**(3):462–70.
70. Keane MM, Gabrieli JD, Mapstone HC, Johnson KA, Corkin S. Double dissociation of memory capacities after bilateral occipital-lobe or medial temporal-lobe lesions. *Brain* 1995;**118**(Pt 5):1129–48.
71. Dijksterhuis A, Bos MW, Nordgren LF, van Baaren RB. On making the right choice: the deliberation-without-attention effect. *Science* 2006;**311**(5763):1005–7.
72. Shanks DR, Newell BR, Lee EH, Balakrishnan D, Ekelund L, Cenac Z, et al. Priming intelligent behavior: an elusive phenomenon. *PLoS One* 2013;**8**(4):e56515.
73. Broadbent DE. Levels, hierarchies, and the locus of control. *Q J Exp Psychol* 1977;**29**:181–201.
74. Bartlett FC. *Remembering: a study in experimental and social psychology*. The Cambridge University Press; 1932.
75. Ballard PB. Oblivescence and reminiscence. *Br J Psychol Monogr* 1913;**Suppl. 1**:1–82.
76. Wheeler MA, Roediger HL. Desperate effects of repeated testing: reconciling Ballard's (1913) and Nartlett's (1932) results. *Psychol Sci* 1992;**3**(4):240–5.
77. Roediger HL, McDermott KB. Creating false memories: remembering words not presented in lists. *J Exp Psychol: Learn Mem Cogn* 1995;**21**(4):803–14.
78. Holden KJ, French CC. Alien abduction experiences: some clues from neuropsychology and neuropsychiatry. *Cognit Neuropsychiatry* 2002;**7**(3):163–78.
79. Smith M, Pazder L. *Michelle remembers*. Congdon & Lattes: distributed by St. Martin's Press; 1980.
80. Manschreck TC, Maher BA, Milavetz JJ, Ames D, Weisstein CC, Schneyer ML. Semantic priming in thought disordered schizophrenic patients. *Schizophr Res* 1988;**1**(1):61–6.
81. Linscott RJ, Knight RG. Automatic hypermnesia and impaired recollection in schizophrenia. *Neuropsychology* 2001;**15**(4):576–85.

82. Butler PD, Martinez A, Foxe JJ, Kim D, Zemon V, Silipo G, et al. Subcortical visual dysfunction in schizophrenia drives secondary cortical impairments. *Brain* 2007;**130**(Pt 2):417–30.
83. Coleman MJ, Cestnick L, Krastoshevsky O, Krause V, Huang Z, Mendell NR, et al. Schizophrenia patients show deficits in shifts of attention to different levels of global-local stimuli: evidence for magnocellular dysfunction. *Schizophr Bull* 2009;**35**(6):1108–16.
84. Johnson SC, Lowery N, Kohler C, Turetsky BI. Global-local visual processing in schizophrenia: evidence for an early visual processing deficit. *Biol Psychiatry* 2005;**58**(12):937–46.
85. Gonzalez-Hernandez JA, Pita-Alcorta C, Cedeno IR. From genes to brain oscillations: is the visual pathway the epigenetic clue to schizophrenia? *Med Hypotheses* 2006;**66**(2):300–8.
86. Schechter I, Butler PD, Jalbrzikowski M, Pasternak R, Saperstein AM, Javitt DC. A new dimension of sensory dysfunction: stereopsis deficits in schizophrenia. *Biol Psychiatry* 2006;**60**(11):1282–4.
87. McCarley RW, Wible CG, Frumin M, Hirayasu Y, Levitt JJ, Fischer IA, et al. MRI anatomy of schizophrenia. *Biol Psychiatry* 1999;**45**(9):1099–119.
88. Nordahl TE, Kusubov N, Carter C, Salamat S, Cummings AM, O'Shora-Celaya L, et al. Temporal lobe metabolic differences in medication-free outpatients with schizophrenia via the PET-600. *Neuropsychopharmacology* 1996;**15**(6):541–54.
89. Barch DM, Cohen JD, Servan-Schreiber D, Steingard S, Steinhauer SS, van Kammen DP. Semantic priming in schizophrenia: an examination of spreading activation using word pronunciation and multiple SOAs. *J Abnorm Psychol* 1996;**105**(4):592–601.
90. Jacoby LL, Woloshyn V, Kelley C. Becoming famous without being recognized: unconscious influences of memory produced by dividing attention. *J Exp Psychol: General* 1989;**118**(2):115–25.

Chapter 3

Blindsight

The experiments on nonconscious memory discussed in Chapter 2, Nonconscious memory, suggest that not all the information that the brain receives and retains get into our conscious awareness. We become consciously aware of a stimulus only if it is processed in the V3A frontal loop. Perception and its awareness therefore are dissociable. Blindsight is a phenomenon that most convincingly demonstrates that dissociation. In this phenomenon, an individual denies seeing anything consciously but is able to make an accurate "guess" about objects placed in the blind field. It is a phenomenon of perception without awareness.

Blindsight was studied systematically by psychologist Lawrence Weiskrantz (Fig. 3.1). In the 1960s while working at Cambridge University as a research associate Weiskrantz noticed that a monkey named Helen was able to "see" light after surgical removal of her primary visual cortex (area V1). Her vision was diminished but not completely lost.[1] It was a surprise observation because at the time it was believed that vision is completely lost if the primary visual cortex is damaged. The observation generated considerable interest and also skepticism, particularly because the autopsy revealed that a small part of Helen's primary visual cortex was intact. Weiskrantz was convinced that Helen's vision could not have come from a small intact part of the visual cortex. He therefore decided to study this phenomenon in human volunteers.

Around the same time, a German psychologist and physiologist Ernst Poppel at Max Planck Institute observed that war veterans who became "blind" following traumatic injuries looked in the direction of the source when light is flashed in their "blind" fields of vision.[2] Since these veterans denied "seeing" light, Poppel asked them to guess the direction and play a game. To everyone's surprise their guesses were remarkably accurate. It was the first clue that the "blind" field is not really visionless in these individuals.

At the time Poppel was publishing his findings, Weiskrantz met a 26-year-old patient nicknamed DB in London's National Hospital. His right occipital cortex including the primary visual cortex was removed because of an invading benign tumor. DB was totally blind in the left visual field except for a small crescent of fuzzy vision in the left upper quadrant. He was

Neuroscience of the Nonconscious Mind. DOI: https://doi.org/10.1016/B978-0-12-816115-9.00003-6

FIGURE 3.1 Lawrence Weiskrantz (1926–2018). Described and demonstrated blindsight. *Reprinted from http://www.scholarpedia.org/article/User:Larry_Weiskrantz/.*

studied extensively by Weiskrantz and Elizabeth Warrington, a neuropsychologist at the National Hospital.[3] Using Popper's technique, he was asked to extend his hand toward the source of a flash of light shown in his "blind" field of vision. DB insisted that he could not do it. He was then asked to guess. His guesses were remarkably accurate. He participated in several other experiments and "guessed" the shape, color, and movement of objects accurately. DB himself was surprised by his performance but continued to maintain that he was just guessing. These experiments revealed that DB could discriminate the shape and movement of objects and correctly guess the direction of the source of light. His ability to discriminate fine lines was not as good as someone with normal vision, but it was remarkably accurate for a "blind" person. Over the next few years, DB began to "feel" things in the blind field but was not able to "see." He would say, "I know there is something but I cannot see." This phenomenon is called type 2 blindsight, as opposed to type 1 or simple blindsight in which individuals do not feel anything in the blind field.

People with type 2 blindsight "feel" the presence of an object, but they cannot describe the feeling. It may be similar to the "gut feeling" we all experience once in a while. However, there is a difference. We are never sure about the accuracy of the gut feelings while individuals with type 2 blindsight are more certain. It is a feeling that cannot be described in terms of known sensory experiences. More significantly, the "feeling" indicates that we perceive more than what we consciously know. It is also possible that some of our decisions are based on stimuli/information that we are not consciously aware of. This possibility was illustrated in experiments Weiskrantz conducted on another individual with blindsight called GY. Since his left visual cortex was damaged in an accident, he was blind in the right hemifield. He had a small zone (<3 degrees) of vision.[4] Like DB, he could discriminate intensity of light sources placed in the blind field. In an interesting experiment, GY was asked to walk in a room using the blind eye. He was able to walk without a problem, avoiding pieces of furniture on his way. When asked why he did not walk in a straight line, he replied "I don't know; I feel like walking that way." Obviously, he was able to "see" the obstacles nonconsciously. This experiment demonstrates how nonconscious information affects our conscious actions and decisions. It opens up the possibility that many of our conscious actions and decisions are guided by nonconscious information we do not know we have! At times, we try to justify those decisions by creating an alibi, just like GY did, but we usually do not know the reasons for our biases, likes, dislikes, and idiosyncratic behaviors. They make us who we are. That is why we do not know real reason for our biases, likings, dislikings and idiosyncratic behaviors. They make us who we are. Those biases could well be because of nonconscious knowledge that the brain has retained without our awareness. Nonconsciously retained information thus defines our personality.

After Weiskrantz's initial study, other investigators began studying people with blindsight. They found that these individuals have ability to discriminate colors, patterns, and movements. They are also known to visually follow moving objects without consciously "seeing" them. Their pupils expand and contract while "looking" at objects and lights, and when asked to grasp an object placed in the blind field, they adjust their fingers and grip to match the shape and size of the objects they cannot see consciously. Interestingly, their reaction time to a flash of light is similar in both their blind and intact fields of vision.[5,6] Therefore, the brain mechanism that controls reaction time remains intact in blindsight.

The ability of individuals with blindsight to perceive stimuli without conscious awareness is similar to subliminal perception, which is also arguably perception without awareness. Both of these phenomena support observations of one of our neuroimaging experiments. In this experiment, we found that conscious perception is not necessary for cognitive processing of a stimulus. Therefore, all stimuli, whether consciously perceived or not, are cognitively processed in the brain.[7,8]

Despite the overwhelming evidence in favor of sight without conscious awareness, skeptics at the time believed that "vision" in blindsight individuals was due to an incomplete lesion of the primary visual cortex. This argument lost much of its weight when blindsight was demonstrated in people who had their entire cerebral hemisphere removed.[9] This demonstration prompted discovery of an alternate visual pathway that blindsight individuals use to "see" objects nonconsciously. Another line of investigation examined why perceived vision does not reach conscious awareness in blindsight.[10]

Before we discuss those studies, let us look at the visual pathway (Fig. 3.2). The retina of each eye sends out about a million nerve fibers to the lateral geniculate body, which is a part of the thalamus. From the lateral geniculate body, most fibers terminate in the primary visual cortex called area V1 (Figs. 2.20, 3.2, and 3.3). From here, these signals are relayed to the visual association areas V1, V2, V3, V3A, V4, and V5 for further processing.

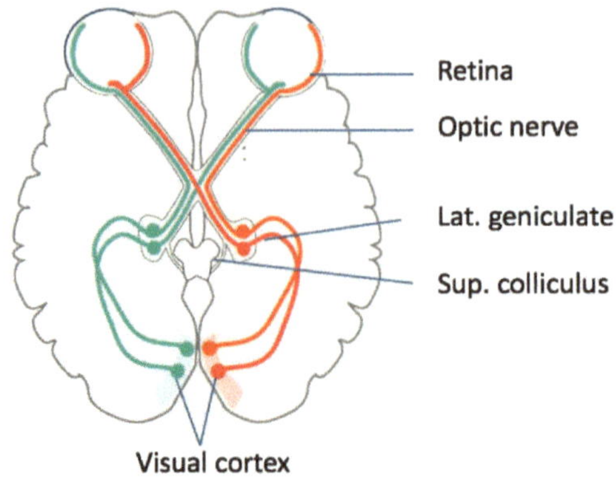

FIGURE 3.2 A schematic diagram of the visual pathway.

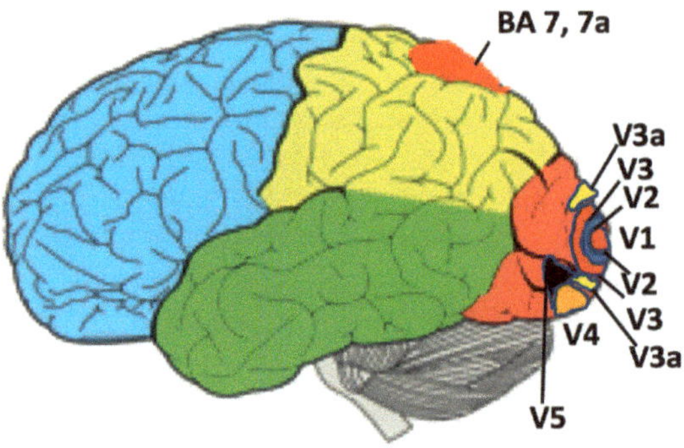

FIGURE 3.3 Locations of visual cortical and parietal areas (BA 7 and 7a) involved in visual perception. Areas V5, V3, V3A, and BA 7 are activated in blindsight. *Image modified from: http://thebrain.mcgill.ca.*

In addition to the lateral geniculate body, about a dozen other brain areas receive visual fibers from the retina, the most prominent among these areas is the superior colliculus, which is located in the midbrain. It receives about 150,000 fibers from the retina.

Fibers that relay in the lateral geniculate body carry information about properties of the visual object, like color, shape, and orientation, while those that relay in the other brain areas carry location and movement signals. Fibers that allow nonconscious perception in blindsight individuals originate in the retina and relay in the superior colliculus and also in the pulvinar, which is a part of the thalamus. These fibers then terminate in visual association areas V2, V3, V3A, V4, and V5 (Figs. 2.20 and 3.3). Visual association areas also receive visual signals directly from the lateral geniculate body.[11] Because of this connection, blindsight individuals can perceive colors, shapes, and orientations of objects in their blind fields. It appears that the visual fibers from the superior colliculus, pulvinar, and lateral geniculate body that terminate in the visual association areas do not activate the brain network that brings visual signals to conscious awareness.

Neuroimaging studies on people with blindsight have found activity in many brain areas (Fig. 3.3) including areas V5, V3, and V3A and BA 7 located in the parietal cortex.[11,12]

Blindsight therefore is dependent on the visual signals carried to the visual association areas directly from the superior colliculus and lateral geniculate body. That is why if the lateral geniculate body is damaged (along with area V1) blindsight is lost and the individual becomes completely blind with no perception with or without conscious awareness.[11] Normal function of the direct connection to the visual association areas from the lateral geniculate body and superior colliculi is to orient our attention to unusual visual events like a flash of light. It initiates the reflex action to draw attention to potentially harmful stimuli/events. These signals do not relay in the primary visual cortex V1 and thus this action is carried out nonconsciously. We become aware of those stimuli only when signals are relayed in area V1 and processed for conscious awareness. This processing is not as quick as nonconscious processing, which is why we react first and notice the source of the flash of light or loud noise later. Signals that are consciously processed carry detailed characteristics of the stimuli, needed to understand their significance. These signals are processed in the V3A frontal loop as discussed in Chapter 2, Nonconscious memory. Activation of this loop makes us consciously aware of a perceived stimulus.[7,8,13,14]

It appears that the visual signals that arrive at area V3A from areas other than the primary visual cortex cannot activate the V3A frontal loop and thus cannot be consciously perceived. Only signals relayed in the primary visual cortex or area V1 can activate this loop. That is why when the primary visual cortex is damaged (as in individuals with blindsight) area V3A does not receive signals that can activate the V3A frontal loop needed for conscious perception of visual stimuli. As noted earlier, in blindsight, area V3A receives visual signals directly from the superior colliculus and lateral geniculate body, but these signals cannot activate the V3A frontal loop and thus cannot be consciously perceived. Despite being perceived nonconsciously, these signals are processed cognitively to allow people with blindsight avoid obstacles and "feel" the presence of objects in their blind fields.

The white matter tract that comprises the V3A frontal loop is not known, but there are two major pathways that connect the visual cortex and frontal lobe. These pathways were studied extensively by neuropsychologists Leslie Ungerleider and Mortimer Mishkin in monkeys.[15] Later, neuropsychologists Branda Milner (Fig. 2.7) and Melvyn Goodale studied them in a patient, called DF, who could see objects but was not able to recognize them (visual agnosia).[16] The two major pathways that connect the visual and frontal cortex are called dorsal and ventral pathways because of their respective positions. The ventral pathway originates in the visual cortex and passes through the medial temporal lobe (MTL) and limbic system, terminating in the frontal cortex. It carries information that helps us recognize objects in the visual field. Because of its connection with the MTL and limbic system, it uses past memory and emotion to recognize and provide historical perspectives on an object. This pathway is classically called the "what" pathway for obvious reasons. The dorsal pathway connects the visual cortex to the frontal lobe via the parietal lobe. It carries information about the location of an object in the visual field, therefore, it is also called the "where" pathway. The information carried in this pathway is object-centric, while it is viewer-centric in the ventral pathway. The two pathways are extensively interconnected.

Even though not verified experimentally, it appears that the V3A frontal loop uses both the dorsal and ventral pathways to connect area V3A and the frontal cortex. Most of its fibers probably use the dorsal route because lesions in the parietal cortex through which the dorsal pathway passes result in impaired awareness of perceived stimuli. Lesions in the posterior part of the parietal causes a condition called hemineglect, discussed in Chapter 4. In this condition a person can see everything but neglects stimuli in the affected field of visions. These people are known to shave half of their faces and draw pictures of only one half of a clock or a visual field. This condition could theoretically be caused by a lesion in the parietal cortex that interrupts the V3A frontal loop, leading to impaired awareness of stimuli. Because of similar neural damage, blindsight and hemineglect individuals share a common feature. In both conditions, there is nonconscious knowledge of stimuli in the blind or neglected field but existence of those stimuli is not consciously acknowledged.

Blindsighted individuals cannot consciously see the stimuli that are perceived nonconsciously because signals arriving directly to area V3A from the superior colliculus and lateral geniculate body cannot activate the V3A frontal loop. It is therefore, at least theoretically possible to restore conscious vision if these visual signals are made to activate the V3A frontal loop in blindsighted individuals. Even though it is a challenge to make that happen, it is not impossible. It can be done with currently available surgical and neuromodulation techniques. Deep brain stimulation, transcranial magnetic stimulation, and emerging neuromodulation techniques could be useful in this context. We discussed this possibility in an earlier publication.[17] Development of such a technique could restore vision in people with blindsight

Blindsight is an example of complete dissociation between perception and awareness, but there are conditions in which the dissociation is incomplete. In these conditions, individuals consciously "know" presence of an object in the visual field but fail to understand its significance. Thus, individuals with one of many "agnostic" conditions can perceive stimuli but do not know what those stimuli mean. Stimuli are not processed to the level that would allow recognition and understanding of their functional significance.

Based on the deficit, there are many types of agnosias. The most common are visual and speech agnosias. In visual agnosia, a person can see an object but cannot identify it; in speech agnosia a person can hear voices but cannot understand meaning. There are many subcategories of each of these agnosias. For example, in a relatively common form of visual agnosia people have difficulty recognizing faces (prosopagnosia) or movement of an object (akinetopsia). Lesions in different brain areas cause different forms of agnosia, but most of them have lesions in the parietal and temporal lobes or at the junction of the occipital, parietal, and temporal lobes (OPT junction).

A different kind of condition in which an individual cannot use perceived information is observed in people with so-called split brain. In these individuals the right and left hemispheres of the brain are separated and the two halves do not communicate. Information received in a hemisphere is processed in the same hemisphere and stays there. As a result, other half of the brain remains unaware of the information.[18,19] Since the two hemispheres are not functionally identical, a stimulus cannot be fully processed unless both hemispheres are involved. That is why split brain leads to cognitive and behavioral impairments. Sometimes each hemisphere receives different stimuli or processes a stimulus differently, leading to dissociation of action, expression, and thinking.

Initial experiments on split brain were conducted by an American neurobiologist named Roger Sperry (Fig. 3.4) who was awarded the Nobel Prize in 1981 for this work. Sperry's interest in psychological processes developed by chance when he came in contact with a psychology professor named R.H. Stetson who was a student of the influential American psychologist William James. Since Stetson was physically handicap, Sperry used to drive him around and attended meetings that Stetson attended. In these meetings, he developed interest in psychology. After training in psychology as a graduate student he received postdoctoral training at Harvard University in the laboratory of Karl Lashley, a well-known psychologist who made significant contributions to our understanding of learning and memory.

FIGURE 3.4 Roger Wolcott Sperry (1913–1994). Awarded the Nobel prize for the study of split brain individuals. *Image reprinted from Archives of California Institute of Technology.*

In Lashley's laboratory, Sperry began studying the brain mechanisms of learning. He was particularly interested in studying processes that are "hardwired" and remain stable throughout life. An example of one such process is moving a limb away from fire. This behavior remains stable and neural circuits involved in this action do not change. Since changes in circuits are essential for learning to occur, hardwired circuits resist relearning. To understand what makes a circuit hardwired, Sperry connected the nerves of the left leg of a rat to the right and found that the animal would lift the right leg when the left was shocked.[20] This response remained unchanged even after several repetitions. It suggested that protective responses are hardwired and cannot be relearned.

In another experiment, he extended the work of Ronald E. Myers of the University of Chicago, dissected the optic nerves of a cat and connected the right eye to the right hemisphere and the left eye to the left hemisphere. Normally, most of the fibers of the optic nerve cross over to the other hemisphere (Fig. 3.2). He also divided the corpus callosum, which is a track of fibers that connect the two hemispheres of the brain (Fig. 3.5). After the surgery, this split brain cat had no apparent disability. This led neurosurgeons to begin cutting the corpus callosum of epileptic patients to control spread of seizures. However, Sperry eventually saw a problem in the cat he was studying. He trained the animal to respond to a triangle while the left eye was open and the right eye was closed. After the cat learned this task, the left eye was closed and made to perform the task with just the right eye open. The cat could not perform it even though the task was learned to perfection over several months with the left eye open.[21] This experiment suggested that each hemisphere processes information based on the task it has learned. Learning does not transfer from one hemisphere to the other in split brain animals. With this knowledge, he started working on people who had their corpus callosum dissected to control seizures.[19]

In another experiment, Sperry presented different objects in the right and left visual fields and asked a split brain volunteer to draw a picture of the object using his left hand. He drew objects presented only on the left visual field because the left hand was controlled by the right hemisphere which can

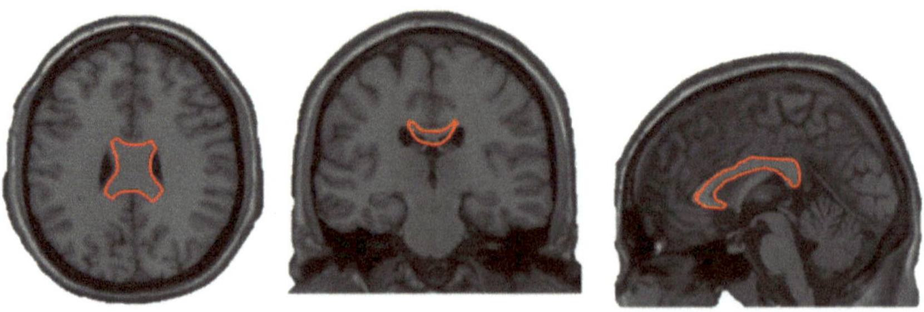

FIGURE 3.5 Images of the human brain (in sagittal, coronal, and transverse views) showing the corpus callosum, which is a track of nerve fibers connecting the two hemispheres of brain.

see objects only on the left visual field. Since the volunteer had to use his left hand, the right hemisphere guided his actions. However, when the same volunteer was asked to describe objects verbally, he described those in the right visual field because the speech center or Broca's area is located in the left frontal cortex (Fig. 3.6), which was able to see objects in the right field. In a similar experiment, Sperry's then graduate student and current professor of psychology at the University of California Santa Barbara, Michael Gazzaniga, asked a 13-year-old split brain patient called PS to name his girlfriend by flashing the question in the left field so that it was perceived in the right hemisphere. He could not say her name because the speech center located in the left hemisphere did not see the question, but when he was asked to write her name with his left hand using scribbles, he immediately wrote three letters, L-I-Z.[22] He was able to do it because the right hemisphere that controlled his left hand perceived and understood the question.

In another interesting experiment, Sperry showed picture of a pencil to a split brain patient in the left field so that it was seen in the right hemisphere. When asked what he saw, the patient said "nothing" because the left hemisphere where the speech center is located had not seen anything. But when asked to choose the object shown in the picture using his left hand he would pick up a pencil from a collection of different objects. When asked why he chose the pencil he did not know. Even though the right hemisphere cannot talk, it can feel and has emotions. However, these emotions cannot be verbalized unless transferred to the left hemisphere where the speech center is located. This dilemma of unverbalized emotion was demonstrated in another experiment on a split brain woman. In this experiment, Sperry and Gazzaniga showed a nude picture in her left hemisphere. She laughed and identified it as a nude picture, but when it was shown in her right hemisphere, she denied seeing anything but a sly smile spread over her face and

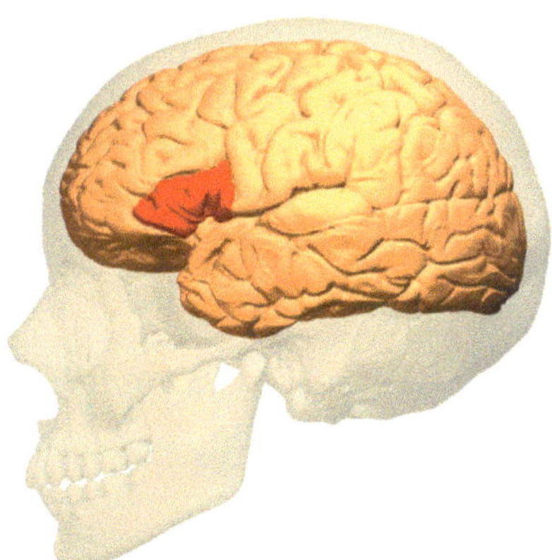

FIGURE 3.6 Broca's area (*red*) located in the left hemisphere is responsible for speech. *Modified from Wikimedia Commons.*

she began to chuckle. When asked why she was smiling, she did not know but thought the machine showing the picture was funny.[18] It is not clear what emotion her right hemisphere experienced. It is also unclear whether she was consciously aware of seeing the nude picture or was cognizance of its cultural and moral significance. It is difficult to predict her opinion of this experience. She probably had a split opinion. The right hemisphere that saw the picture could have had one opinion and the left may have had a different opinion because it never saw the picture. Since the left hemisphere can talk, she could talk about not seeing the picture, but if she was asked to write using her left hand, she would express a different opinion because the right hemisphere which has seen the picture will control that opinion. Moreover, she would have two different opinions if asked to write with the right or left hand.

In split brain individuals the left hemisphere does not know what the right is doing or thinking. Therefore, sometimes the left hand will do a task that the right would undo. In fact, one man said he would button his shirt with his left hand and then his right hand would unbutton it.[23] People have also reported one hand picking up a grocery item and putting it in the cart and the other hand removing it and putting in another item. In extreme cases of split brain different halves of the brain show different likes and dislikes.

This leads to an interesting question: what would happen if there was a partial split and information was only partially exchanged between the two hemispheres? We do not know much about such a scenario, but it is entirely possible that many of us or possibly all of us have some degree of partial structural or functional split between the two hemispheres. Damage to just a few fibers of the corpus callosum (or anterior and posterior commissures that are also fibrous connections between the two hemispheres) by microemboli or microscopic hypoxia could cause significant deficits in inter-hemispheric information processing. Could this be the reason we are sometimes confused about our own intentions and behaviors? If this confusion becomes pathological, a diagnosable psychiatric condition could result. Partial split brain could thus be an unstudied cause of psychiatric illness.

Bibliography

1. Crowey A, Weiskrantz L. A perimetric study of visual field defects in monkeys. *Quart J Exp Psychol* 1963;**15**:91–115.
2. Poppel E, Held R, Frost D. Residual visual function after brain wounds involving the central visual pathways in man. *Nature* 1973;**243**(405):295–6.
3. Weiskrantz L. *Blindsight: a case study and implications*. Clarendon Press; 1986.
4. Weiskrantz L, Barbur JL, Sahraie A. Parameters affecting conscious versus unconscious visual discrimination with damage to the visual cortex (V1). *Proc Natl Acad Sci USA* 1995;**92**(13):6122–6.
5. Ter Braak JWG, Schenk VWD, Van Vliet GM. Visual reactions in a case of long-lasting cortical blindness. *J Neurol Neurosurg Psychiatry* 1971;**34**(2):140–7.

6. Marcel AJ. Conscious and unconscious perception: an approach to the relations between phenomenal experience and perceptual processes. *Cognit Psychol* 1983;**15**(2):238–300.
7. Badgaiyan RD. Cortical activation elicited by unrecognized stimuli. *Behav Brain Funct* 2006;**2**(17):1–5.
8. Badgaiyan RD. Nonconscious perception, conscious awareness and attention. *Conscious Cogn* 2012;**21**(1):584–6.
9. Ptito A, Fortin A, Ptito M. 'Seeing' in the blind hemifield following hemispherectomy. *Prog Brain Res* 2001;**134**:367–78.
10. Leh SE, Johansen-Berg H, Ptito A. Unconscious vision: new insights into the neuronal correlate of blindsight using diffusion tractography. *Brain* 2006;**129**(Pt 7):1822–32.
11. Schmid MC, Mrowka SW, Turchi J, Saunders RC, Wilke M, Peters AJ, et al. Blindsight depends on the lateral geniculate nucleus. *Nature* 2010;**466**(7304):373–7.
12. Goebel R, Muckli L, Zanella FE, Singer W, Stoerig P. Sustained extrastriate cortical activation without visual awareness revealed by fMRI studies of hemianopic patients. *Vision Res* 2001;**41**(10–11):1459–74.
13. Badgaiyan RD. Nonconscious processing, anterior cingulate, and catatonia. *Behav Brain Sci* 2002;**25**:578–9.
14. Badgaiyan RD. Conscious awareness of retrieval: an exploration of the cortical connectivity. *Int J Psychophysiol* 2005;**55**(2):257–62.
15. Ungerleider LG, Mishkin M. Two visual systems. In: Ingle DJ, Goodale MA, Mansfield RJW, editors. *Analysis of visual behavior*. MIT Press; 1982. p. 549–86.
16. Goodale MA, Milner AD, Jakobson LS, Carey DP. A neurological dissociation between perceiving objects and grasping them. *Nature* 1991;**349**(6305):154–6.
17. Badgaiyan RD. Manipulation of the extrastriate frontal loop can resolve visual disability in blindsight patients. *Med Hypotheses* 2012;**79**(6):767–9.
18. Gazzaniga MS. The split brain in man. *Sci Am* 1967;**217**(2):24–9.
19. Sperry RW. Cerebral organization and behavior: the split brain behaves in many respects like two separate brains, providing new research possibilities. *Science* 1961;**133**(3466):1749–57.
20. Weiss P, Sperry RW. Unmodifiability of muscular coordination in the rat, demonstrated by uscle transposition and nerve crossing. *Am J Physiol* 1940;**129**:492.
21. Myers RE, Sperry RW. Interocular transfer of a visual form discrimination habit in cats after section of the optic chaism and corpus callosum. *Anatomical Records* 1953;**115**:351–2.
22. LeDoux JE, Wilson DH, Gazzaniga MS. A divided mind: observations on the conscious properties of the separated hemispheres. *Ann Neurol* 1977;**2**(5):417–21.
23. Bogen GE. The calosal syndrome. In: Heilman KM, Valenstein E, editors. *Clinical neuropsychology*. Oxford University Press; 1979. p. 308–58.

Chapter 4

Hemineglect

Hemineglect (also called unilateral or hemispatial neglect), like blindsight, is a relatively recent discovery. It is an interesting condition in which an individual ignores half of consciously perceived stimuli. For example, people with visual hemineglect ignore everything they see in the left visual field. This condition was first described in 1941 by a British neurologist named Walter Russell Brain (Fig. 4.1). He observed that three of his patients who had lesions in the posterior part of the right cortex exhibited strange behavior would get disoriented in their own homes. Brain correctly interpreted their problem as "agnosia for the left half of external space."[1] Brain was a well-known neurologist at Oxford University and his book *Brain's Diseases of the Nervous System*[2] is still used as a comprehensive neurology textbook. He was editor of the journal "Brain" for several years and was involved in the treatment of the then British Prime Minister Winston Churchill at the end of his life in 1965.

After Brain described this condition, other physicians reported similar symptoms in their patients. It turned out to be a relatively common condition in people with stroke. Approximately 80% of individuals with lesions in the right posterior parietal cortex show signs of neglect for stimuli in the contralateral visual field. They ignore objects on the left side; some of them even ignore the left side of their bodies, while having intact visual and sensory perception. They are known to shave one half of their face, put on makeup on one side, and not eat food on half of the plate even when hungry. If asked to draw a picture of the surroundings they would ignore details on the left side, and draw a clock face either by using only numbers 1–6 or writing all 12 numbers on the right side (Fig. 4.2). In some cases, they even deny the existence of left side of their own bodies. In extreme cases, called somatoparaphrenia, affected people think that their limbs on the affected side belong to someone else. Oliver Sacks, a British neurologist and famed author of several popular books and movies, described an interesting case. One of his patients had disowned the left part of the body to the extent that once he fell out of bed trying to get rid of "someone else's leg" inside his blanket.[3]

Neglect in these individuals is not limited to vision. They tend to ignore auditory, olfactory, proprioceptive, and tactile stimuli also on the affected side. Even their memory of objects on the left side is compromised; while

Neuroscience of the Nonconscious Mind. DOI: https://doi.org/10.1016/B978-0-12-816115-9.00004-8

FIGURE 4.1 Walter Russell Brain (1895–1966). Described hemineglect in 1941. *Reprinted from Wikimedia Commons.*

recalling a scene, they usually recall objects located only on the healthy side. However, it is not clear whether their inability to recall scenes on the neglected side is due to impaired memory or because they did not attend to objects on that side in the first place. As discussed in the previous chapter, if objects are not attended, memory for that object is impaired.[4]

Based on the nature of neglect there are two primary types of hemineglect: egocentric and allocentric. Individuals with egocentric neglect tend to ignore everything that falls on the neglected side (left as a result of the lesions on the right). They have no problem processing images in the right visual field but will ignore everything on the left. In contrast to this, people with allocentric neglect, ignore one side of each object in both visual fields. Look at the image at the bottom of Fig. 4.2. A person with allocentric neglect drew only the right side of the house, neglecting the left half. In egocentric neglect, the person will either draw the whole house if it is in the healthy field or will not draw anything if it is presented in the neglected field.

There are many subtypes of hemineglect based on the stimuli neglected (e.g., visual, auditory, somatic, etc.). Some of these individuals are unable to recall dreams or imagine objects in the neglected hemifield.[5] This condition is called representational neglect. These individuals also have difficulty performing motor movements. Based on their limited motor skills, neglect can either be motor or premotor. In motor neglect they cannot move limbs of the affected side and in premotor neglect they cannot move the limbs of the unaffected side to the affected side.[6] Some hemineglect individuals while describing a route disregard turns on the neglected side and remember only those on the healthy side. Interestingly many of them do not even know that they have a deficit. This condition is called anosognosia.

Hemineglect is observed mostly in people with lesions in the right posterior parietal lobe (Fig. 4.3).[7] A lesion in the left parietal rarely causes neglect because the right hemisphere controls the space of both sides. The left parietal controls space of only the right side. That is why lesions on the right parietal impair the field that is not controlled by the left parietal. Hemineglect is also reported following lesions of the right frontal lobe even

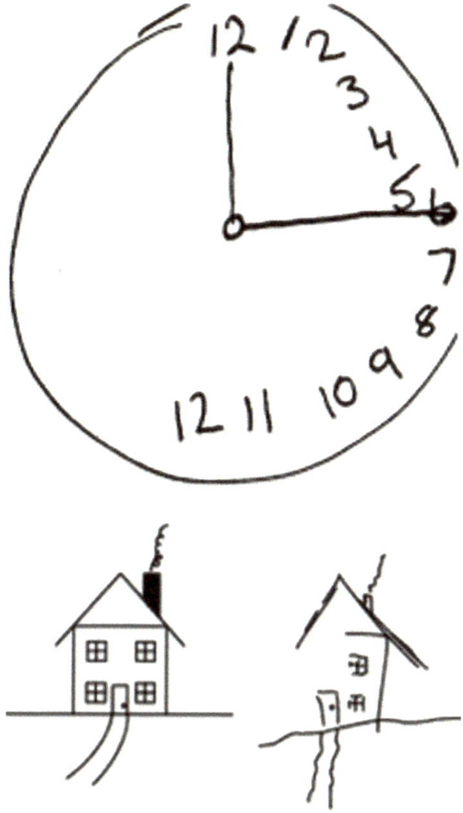

FIGURE 4.2 A clock face drawn by a hemineglect individual (top). When asked to copy a picture of a house (bottom left), another hemineglect individual neglected to draw left half of the picture (bottom right). *Reproduced from Wikimedia Commons.*

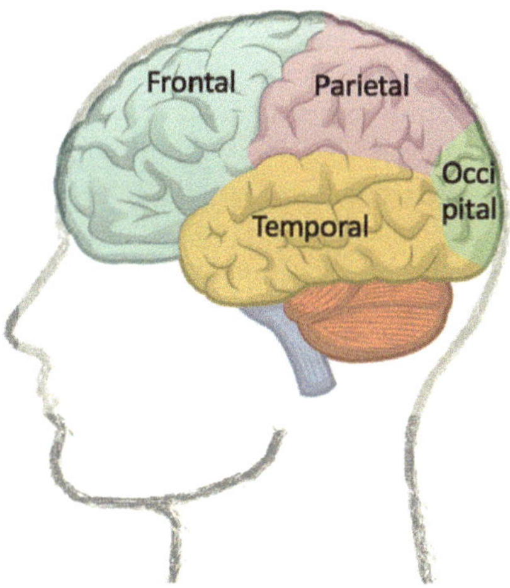

FIGURE 4.3 Hemineglect is relatively common following a lesion in the right posterior parietal lobe. The figure shows relative position of the parietal lobe along with the frontal, temporal, and occipital lobes. *Modified from Wikimedia Commons.*

when the parietal lobe is intact.[8] These individuals usually have allocentric neglect in which one half of each object in either field is neglected. We do not know the neural basis of this kind of neglect, and it is not clear why only one half of an object is neglected.

Hemineglect individuals show another intriguing unexplained phenomenon. When their visual field is reversed using a mirror, they neglect the side that falls on the neglected side of the unreversed image even though the side is reversed. This defies logic and is not consistent with any known processing mechanism of the nervous system.

Another interesting aspect of hemineglect is perception and processing of neglected stimuli. Experiments have demonstrated that despite not being consciously receognized stimuli on the neglected side are perceived and processed for their meaning nonconsciously. Nonconscious perception of neglected stimuli was experimentally demonstrated using an implicit memory task by John Marshall and Peter Halligan of Cardiff University.[9] In this experiment a hemineglect patient known as PS was shown pictures of two houses; one on the healthy and the other on the left side. When asked to tell the difference between the two pictures, PS responded, "they are identical." She ignored the fire in the image in bright red coming out of the house on the ignored left visual field. Even when prompted to look carefully at the left side of the picture, she did not notice the fire. However, when she was asked which house she would prefer to live in, she thought it was a trick question because for her both houses were identical. But when forced to respond, in 9 out of 11 trials she chose the house that was not on fire.[9] Even though consciously she did not "see" the fire PS was nonconsciously aware of it and based on that awareness she preferred the intact house. In another experiment a hemineglect volunteer was given a semantic priming task. When a word was shown on the neglected side his response to a semantically related word shown on the healthy side was significantly quicker.[10] For example, if the word "nurse" was shown on the neglected side, identification of the word "doctor" shown on the healthy side was quicker.

These experiments suggest that individuals with hemineglect not only perceive stimuli on the neglected side but also process them cognitively and understand their meaning. Normal perception of stimuli on the neglected side is also indicated by the observation of similar pattern of activation is observed in this area on the healthy and neglected side.[11,12]

In a way hemineglect is similar to blindsight. In both conditions stimuli are perceived nonconsciously but individuals deny them consciously. It appears that in both of these conditions stimuli do not activate the network that brings them to conscious awareness. As discussed earlier a stimulus cannot be consciously appreciated unless it activates the V3A frontal loop. Even though there are similarities between blindsight and hemineglect, investigators have generally not tied them together for understanding neurocognitive bases of perception without awareness. Each of these phenomenon has been independently investigated and separate hypotheses are proposed concerning neurocognitive deficits. Blindsight is considered a condition of perception without awareness but most researchers do not consider hemineglect as perception without awareness. Hemineglect is considered a deficit of spatial representation or attentional processing.

It is logical to describe hemineglect as a condition in which the attention system fails to orient to the affected side. As discussed earlier, orientation of attention is a necessary component of conscious cognitive processing. It is, however, not required for nonconscious processing.[4] Since the parietal lobe, which is damaged in hemineglect, is associated with orientation, it is possible that the orienting process is impaired in these individuals.[13,14] This possibility is particularly strong because Michael Posner found that individuals with parietal lobe injury have difficulty orienting attention to a target.[15] Because of this difficulty stimuli on the affected hemifield are never consciously processed. The observation that these individuals have an impaired ability to sustain attention also indicates impairment of attentional processing.[16]

The hypothesis that hemineglect is a result of impaired attention processing seems logical, but it does not explain all forms of hemineglect. For example, the attentional system is not known to orient toward only one half of an object (as seen in allocentric neglect). Moreover, it does not explain why a person would deny ownership of body parts. Because attention theory does not explain some of these features, alternate possibilities have been proposed. One of the alternative theories suggest that these individuals have difficulty making spatial representation of space around them. As a result they are unable to make a topographic representation of the neglected field that, for them, does not exist.[17] However, there is no known mechanism that allows the brain to make a topographic map of only one half of an object. It therefore also does not explain allocentric neglect.

Could hemineglect be considered a condition similar to blindsight—perception without awareness? Probably, except in hemineglect loss of awareness is not for just visual stimuli. It could extend to any sensory modality. To a hemineglect individual the world does not exist on the neglected side. Let us see how that might happen. We suggested earlier that conscious awareness of nonconsciously perceived information is brought to conscious awareness when the V3A frontal loop is activated.[18] As discussed in Chapter 3, Blindsight, signals from the visual cortex project to the frontal cortex using two distinct pathways with distinct functions.[19] The pathway that passes through the temporal lobe is called the ventral pathway and the one that passes through the parietal lobe is called the dorsal pathway because of their relative positions.

The concept of two separate visual pathways was first proposed in 1969 by Gerald E. Schneider, a psychologist at Massachusetts Institute of Technology.[20] Even though not all aspects of his proposal survived later scrutiny, the basic concept of separate systems for visual coding of location and object identification is generally accepted with a few modifications. These pathways were studied extensively in monkeys by Mortimer Mishkin and Leslie Ungerleider at the National Institutes of Health. They mapped the two pathways. We now know that the ventral pathway courses ventrally interconnecting visual association areas (V2, V3, V3A, V4, and V5) and inferior temporal cortex. The dorsal pathway also connects the visual association areas but then courses dorsally into the inferior parietal cortex. The ventral pathway enables visual identification of objects ("what" pathway) and the dorsal pathway locates an object in the visual field ("where" pathway).[19] The ventral pathway carries only visual information but the dorsal pathway carries multimodal information from the visual and other sensory systems.[21] It is relayed in area V3A and in the posterior parietal cortex and uses the superior longitudinal fasciculus to terminate in the dorsal prefrontal cortex.[22] Most fibers of the V3A frontal loop probably take the dorsal pathway because it carries spatial information, and also because damage to the parietal cortex causes loss of conscious appreciation of stimuli—hemineglect.

Thus, the neuropathology of blindsight and hemineglect converge at the V3A frontal loop. As discussed in previous chapters, in blindsight visual signals that are relayed in area V3A directly from the lateral geniculate body do not activate the V3A frontal loop and thus are not consciously perceived. In hemineglect, because of injury to the posterior parietal cortex, the dorsal visual pathway responsible for spatial localization may be damaged. The lesion also damages the V3A frontal loop resulting in failure of stimuli to enter conscious awareness, resulting in neglect. Therefore, a lesion in the posterior parietal cortex would affect spatial representation as well as conscious awareness of stimuli, resulting in neglect.

The fact that some hemineglect individuals neglect and disown their own body parts on the affected side is consistent with known functions of the parietal cortex. The posterior parietal is known to use visual signals from the dorsal pathway and inputs from other brain areas to make a map of the "self." Thereafter this "self" establishes ownership of body parts using signals from posterior cingulate and medial temporal lobe. A lesion in any of these areas therefore interferes with establishment of ownership[23] and distort the spatial map of the "self."[24] But the meaning of the term self is somewhat vague both in philosophy and neurobiology.[25] However, it is closely tied to the concept of consciousness discussed in Chapter 1, Historical perspective. A detailed discussion of this subject is beyond the scope of this book but is relevant in the context of hemineglect. The concept of autonoetic consciousness refers to the ability of a person to be "self-reflective" or "self-knowing." It connects a person to his or her past memories and future directions and differentiates "self" from "others."[26,27] Parietal damage possibly results in loss of autonoetic consciousness, which is why it impairs the ability to distinguish part of the "self" from those of "others."[24]

An important related function of the parietal lobe is to project the visual images constructed in the retina and visual cortex to the outside world. Because of this projection we do not see images on the retina but visualize them as scenes outside the body. The brain mechanism involved in this process is unclear but it has intrigued both neuroscientists and philosophers. In recent years, small but significant discoveries have been made to understand how the brain makes us recognize our position in space. It begins with neurons in the parietal and hippocampal areas that respond to the position of the body and head. These neurons, based on movement of the head, compensate visual scenes. That is why changing the head's position does not alter images of the outside world, even though it alters its orientation on the retina.[28] Head movement is compensated by specialized cells that detect movement and modify images coming to retina. It appears that head movement is referenced to either the body or the world outside by a separate cluster of parietal neurons located in the lateral intraparietal cortex and in Brodmann area 7a, respectively (Fig. 4.4).[29,30] Neurons in these areas are known to use visual and other sensory information to make a composite picture of the world around us. This is accomplished by clusters of neurons, each of which respond to different aspects of movement of the body and the objects around us. These neurons use input from a specialized set of cells called place cells (discussed in the following) located in the hippocampus to analyze visual signals. By integrating these signals, the neurons of the parietal cortex and hippocampus project images formed on the retina to outside of the body. These cells also provide a realistic perspective of the visual field. This perspective helps us understand that objects at a distance look smaller. The neurons of the hippocampus and parietal cortex thus work together to place us in our external environment.[31]

As discussed earlier, the medial temporal lobe is involved in memory processing. In addition, it also maps our location in space as discussed in the previous paragraph. For the discovery of these functions, three scientists were awarded the Nobel Prize in 2014. They are an American-British

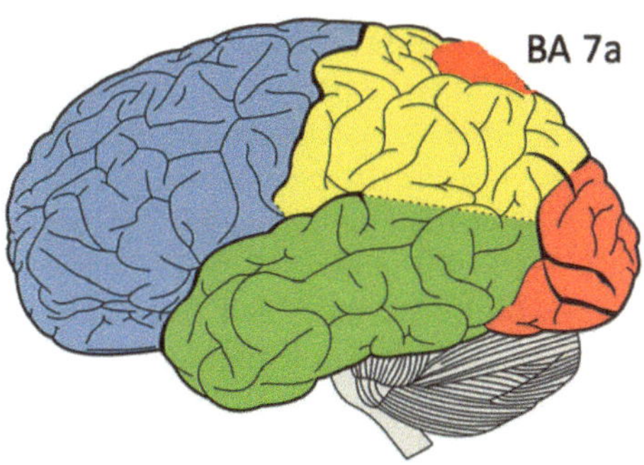

FIGURE 4.4 Location of the Brodmann area (BA) 7a. Neurons in this area locate our position in space. *Modified from http://thebrain.mcgill.ca.*

FIGURE 4.5 John O'Keefe (born 1939). American-British neuroscientist who discovered place cells. *Reproduced from Wikimedia Commons.*

neuropsychologist John O'Keefe (Fig. 4.5) of the University College London; and a then husband-wife team (divorced in 2016) Edward and May-Brit Moser of Norway. In a paper published in 1971, O'Keefe and his student Jonathan Dostoevsky described the hippocampal neurons that fire only when a rat is in a particular location in the cage.[32] When the animal moves, these neurons stop firing and another set of neurons representing the new location become active. These cells are called place cells. Each of these cells represent a specific location in space but some of them, called complex place cells, have more specialized functions. These cells fire if an expected reward at a specific location is not found while others fire when an unexpected novel object is found at a location. Some of these cells are sensitive to changes in environment.[33] Thus, different colors in the surroundings activate different sets of place cells. Changes in smell also alter firing of these cells. Collectively, place cells are responsible for spatial memory. Therefore, they are activated when space-specific memories are retrieved either by mentally tracing a path or by locating an object in a spatial field.

Even though most of the data on place cells were obtained in rats, evidence suggests that these cells exist in the human brain also. Thus, a neuroimaging study conducted on London taxi drivers found significant activation in the hippocampus, indicating increased activity of place cells, when they recalled the route between two points in London.[34]

Place cells get most of their input from other location-sensitive cells called grid cells located close to the hippocampus in the entorhinal cortex.[35,36] These cells were discovered by the Mosers. Like place cells, grid cells also respond to the location but have different characteristics. Their receptive fields form a triangular space, thus the name grid cells. The orientation and size of the field vary but each cell has its own space. These cells characterize different aspects of spatial location. For example, in rats, some of them respond only when stationary while others are activated when moving in the direction of the head or tail. Grid cells do not need auditory or visual cues to locate space and allow animals to move in a familiar space without external cues.

Place and grid cells together make a map of the surroundings as seen by the animal. This map is called an allocentric map. In contrast, cells of the parietal cortex create an egocentric map, which is a map of the location of animal in the surrounding space. Thus, while place and grid cells make a map of the surroundings and place an animal in it, parietal cells identify personal space.[31] This identification is lost in hemineglect. As a result, these individuals disown their own body parts.

It appears that in hemineglect the V3A frontal loop is damaged in the parietal lobe. It passes through the parietal on the way to the frontal cortex. Because of the damage, stimuli on the neglected side do not enter conscious awareness. We discussed earlier that the V3A frontal loop is responsible for bringing a stimulus to conscious awareness. That is the reason damage to the white matter tract between the parietal and frontal cortex is considered the primary cause of hemineglect.[37]

Irrespective of the cause, hemineglect mostly remains undetected. Even though some individuals show signs of neglect in daily life they rarely seek help. That is why it is important to examine these individuals using clinical tests. A number of tests have been developed to detect different types of neglect in clinical settings. For example, a simple test to detect the most common type of hemineglect, visual hemineglect, is the line bisection task. In this task, individuals are asked to mark the midpoint of a line. For a person with visual neglect, the line extending to the neglected side does not exist, therefore he or she marks the midpoint of the part that falls only in the healthy field. Another commonly used test is a copying task that requires copying one or more drawings that extend to both, the healthy and neglected visual fields.[38] People with the deficit would ignore the part that falls on the neglected side.

Despite being a relatively minor deficit, hemineglect can have serious consequences if these individuals ignore obstruction on neglected field and fall. Similarly, there can be serious consequences of not recognizing one's body parts. People have tried to sever limbs that they thought were alien. Because of these and other problems, most people with hemineglect are not able to live independently.

Unfortunately, at this time there is no definitive treatment for this condition even though a number of therapeutic strategies have been attempted. Some of these strategies have achieved limited success. These strategies include the use of prisms to deviate images of the neglected side to the healthy side. This allows the person to acknowledge the whole visual field.[39] Sometimes forcing individuals to directly gaze toward the neglected side with or without a prism works. Relatively better and more predictable results are reported by activating the healthy side of the brain using repetitive transcranial magnetic stimulation, but this provides only temporary relief.[40] A number of pharmacological agents also have limited therapeutic effect on some individuals. These agents include dopaminergic (levodopa, bromocriptine)[41] and apha-2 noradrenergic (guanfacine)[42] drugs.

Bibliography

1. Brain WR. A form of visual disorientation resulting from lesions of the right cerebral hemisphere. *Proc R Soc Med* 1941;**34**(12):771–6.
2. Brain WRB, Donaghy M. *Brain's diseases of the nervous system*. 12th ed. Oxford University Press; 2009.
3. Sacks O. *The man who mistook his wife for a hat and other clinical tales*. 1st Perennial Library ed. Perennial Library; 1987.
4. Badgaiyan RD. Nonconscious perception, conscious awareness and attention. *Conscious Cogn* 2012;**21**(1):584–6.
5. Coslett HB. Neglect in vision and visual imagery: a double dissociation. *Brain* 1997;**120**(Pt 7):1163–71.
6. Heilman KM, Bowers D, Coslett HB, Whelan H, Watson RT. Directional hypokinesia: prolonged reaction times for leftward movements in patients with right hemisphere lesions and neglect. *Neurology* 1985;**35**(6):855–9.
7. Vallar G. Spatial hemineglect in humans. *Trends Cogn Sci* 1998;**2**(3):87–97.
8. Husain M, Kennard C. Visual neglect associated with frontal lobe infarction. *J Neurol* 1996;**243**(9):652–7.
9. Marshall JC, Halligan PW. Blindsight and insight in visuo-spatial neglect. *Nature* 1988;**336** (6201):766–7.
10. McGlinchey-Berroth R, Milberg WP, Verfaellie M, Alexander M, Kilduff P. Semantic priming in the neglected field: evidence from a lexical decision task. *Cogn Neuropsychol* 1993;**10**:79–108.
11. Rees G, Wojciulik E, Clarke K, Husain M, Frith C, Driver J. Unconscious activation of visual cortex in the damaged right hemisphere of a parietal patient with extinction. *Brain* 2000;**123**(Pt 8):1624–33.
12. Driver J, Vuilleumier P, Eimer M, Rees G. Functional magnetic resonance imaging and evoked potential correlates of conscious and unconscious vision in parietal extinction patients. *Neuroimage* 2001;**14**(1 Pt 2):S68–75.
13. Posner M, Badgaiyan R. Attention and neural networks. In: Parks RW, Levine DS, editors. *Fundamentals of neural network modeling*. The MIT Press; 1998. p. 61–76.
14. Bisley JW, Mirpour K, Arcizet F, Ong WS. The role of the lateral intraparietal area in orienting attention and its implications for visual search. *Eur J Neurosci* 2011;**33** (11):1982–90.
15. Posner MI, Walker JA, Friedrich FJ, Rafal RD. Effects of parietal lobe injury on covert orienting of visual attention. *J Neurosci* 1984;**4**:1863–74.
16. Robertson IH. Do we need the "lateral" in unilateral neglect? Spatially nonselective attention deficits in unilateral neglect and their implications for rehabilitation. *Neuroimage* 2001;**14**(1 Pt 2):S85–90.
17. Bisiach E, Luzzatti C. Unilateral neglect of representational space. *Cortex* 1978;**14**:129–33.
18. Badgaiyan RD. Conscious awareness of retrieval: an exploration of the cortical connectivity. *Int J Psychophysiol* 2005;**55**(2):257–62.
19. Ungerleider LG, Mishkin M. Two visual systems. In: Ingle DJ, Goodale MA, Mansfield RJW, editors. *Analysis of visual behavior*. MIT Press; 1982. p. 549–86.
20. Schneider GE. Two visual systems. *Science* 1969;**163**(3870):895–902.
21. Mishkin M, Ungerleider LG. Contribution of striate inputs to the visuospatial functions of parieto-preoccipital cortex in monkeys. *Behav Brain Res* 1982;**6**(1):57–77.

22. Ungerleider LG, Galkin TW, Desimone R, Gattass R. Cortical connections of area V4 in the macaque. *Cereb Cortex* 2008;**18**(3):477–99.
23. Guterstam A, Bjornsdotter M, Gentile G, Ehrsson HH. Posterior cingulate cortex integrates the senses of self-location and body ownership. *Curr Biol* 2015;**25**(11):1416–25.
24. Dolins FL, Mitchell RW. *Spatial cognition, spatial perception: mapping the self and space*. Cambridge University Press; 2010.
25. Levin JD. *Theories of the self*. Taylor and Francis; 1992.
26. Posner MI, Rothbart MK. Attention, self-regulation and consciousness. *Philos Trans R Soc B: Biol Sci* 1998;**353**(1377):1915–27.
27. Gardiner DM, Raigrodski AJ. Psychosocial issues in women's oral health. *Dent Clin North Am* 2001;**45**(3):479–90 vi.
28. Brotchie PR, Andersen RA, Snyder LH, Goodman SJ. Head position signals used by parietal neurons to encode locations of visual stimuli. *Nature* 1995;**375**(6528):232–5.
29. Sakata H, Shibutani H, Kawano K, Harrington TL. Neural mechanisms of space vision in the parietal association cortex of the monkey. *Vision Res* 1985;**25**(3):453–63.
30. Zipser D, Andersen RA. A back-propagation programmed network that simulates response properties of a subset of posterior parietal neurons. *Nature* 1989;**1988**(331):679–84.
31. Baumann O, Mattingley JB. Dissociable roles of the hippocampus and parietal cortex in processing of coordinate and categorical spatial information. *Front Human Neurosci* 2014;**8**:73.
32. O'Keefe J, Dostrovsky J. The hippocampus as a spatial map. Preliminary evidence from unit activity in the freely-moving rat. *Brain Res* 1971;**34**(1):171–5.
33. O'Keefe J, Nadel L. *The hippocampus as a cognitive map*. Clarendon Press; Oxford University Press; 1978.
34. Maguire EA, Frackowiak RSJ, Frith CD. Recalling routes around London: activation of the right hippocampus in taxi drivers. *J Neurosci* 1997;**17**(18):7103–10.
35. Leutgeb JK, Leutgeb S, Moser MB, Moser EI. Pattern separation in the dentate gyrus and CA3 of the hippocampus. *Science* 2007;**315**(5814):961–6.
36. Langston RF, Ainge JA, Couey JJ, Canto CB, Bjerknes TL, Witter MP, et al. Development of the spatial representation system in the rat. *Science* 2010;**328**(5985):1576–80.
37. Bartolomeo P, Thiebaut de Schotten M, Doricchi F. Left unilateral neglect as a disconnection syndrome. *Cereb Cortex* 2007;**17**(11):2479–90.
38. Parton A, Malhotra P, Husain M. Hemispatial neglect. *J Neurol Neurosurg Psychiatry* 2004;**75**(1):13–21.
39. Rossetti Y, Rode G, Pisella L, Farne A, Li L, Boisson D, et al. Prism adaptation to a rightward optical deviation rehabilitates left hemispatial neglect. *Nature* 1998;**395**(6698):166–9.
40. Oliveri M, Bisiach E, Brighina F, Piazza A, La Bua V, Buffa D, et al. rTMS of the unaffected hemisphere transiently reduces contralesional visuospatial hemineglect. *Neurology* 2001;**57**(7):1338–40.
41. Fleet WS, Valenstein E, Watson RT, Heilman KM. Dopamine agonist therapy for neglect in humans. *Neurology* 1987;**37**(11):1765–70.
42. Malhotra PA, Parton AD, Greenwood R, Husain M. Noradrenergic modulation of space exploration in visual neglect. *Ann Neurol* 2006;**59**(1):186–90.

Chapter 5

Attention

We all know the meaning of attention. In cognitive terms, it is a process by which multiple brain resources focus on a specific task. This characterization is not fundamentally different from what the famous psychologist William James wrote in his book *Principles of Psychology*[1] published in 1890. He wrote, "attention is taking possession of by the mind, in clear and vivid form, of one out of what seem several simultaneously possible objects or trains of thought. Focalization, concentration of consciousness are of its essence. It implies withdrawal from some things in order to deal effectively with others."

By focusing brain resources on one task we enhance our capabilities. It is therefore a process that makes us accomplish tasks that otherwise may not be possible. It may surprise many that ancient civilization recognized this ability and even used it to enhance physical and mental capabilities thousands of years ago. Increased cognitive and physical power acquired using various forms of martial arts is based on the practitioner's ability to focus and control the attentional system. Martial arts that require attentional resources were developed in China and India over 4000 years ago. In China, it originated during the reign of mythological hero Yellow Emperor Huangdi. He is credited with writing books on astrology, medicine, and martial arts. Around the same time different types of martial art techniques based on attentional control were developed independently in north and south India. Techniques developed in the north are described in the ancient text *Agni Puran* (Book of Fire) and those developed in south India are recorded in another text called *Sangam* (Confluence), which was originally scripted in Tamil between 300 BCE and CE 300. With the spread of Buddhism from India to China, techniques developed in the two countries were mixed, leading to the modern martial art techniques we know today like judo, karate, etc. In these techniques attention is focused both consciously and nonconsciously to enhance physical power. Since nonconscious attention cannot usually be controlled voluntarily, practitioners of these techniques use spiritual routines including meditation to gain control over nonconscious attention. That is why it is customary for practitioners of martial arts to follow strict spiritual regimens.

Neuroscience of the Nonconscious Mind. DOI: https://doi.org/10.1016/B978-0-12-816115-9.00005-X

FIGURE 5.1 Michael I. Posner (born 1936). A pioneer of modern cognitive neuroscience. *Picture courtesy, Prof. Posner.*

The ancient scriptures suggest that attentional resources were used in the past not only to enhance physical strength but also to improve perception and cognition. For example, extraordinary "vision" acquired during meditation, specifically in a state of *samadhi* described in *Patanjali Yoga Sutra*,[2] could apparently be achieved by focused attention.

Contemporary theories of attention do not contradict what ancient philosophers thought and practiced. Along with Michael Posner (Fig. 5.1), I summarized modern concepts in a book chapter published a few years ago.[3] The chapter includes some of the highly influential experiments conducted in Posner's laboratory at the University of Oregon. Michael Posner is a pioneer of modern cognitive neuroscience, and the most influential researcher on attention. In the chapter, we discuss three major functions of the attentional system: orienting to sensory stimuli, control of cognitive function, and maintenance of alert state.

As discussed in previous chapters, attention is necessary for a stimulus to become consciously available. Unless attention is oriented to it, a stimulus cannot be consciously processed or perceived.[4] While all conscious processes activate attentional system, activation of the system does not necessarily mean that the process is conscious. It can be activated by both conscious as well as nonconscious processes.

There are two forms of attention, overt and covert. Broadly speaking (but not exactly as discussed later), overt attention can be called conscious attention, and covert nonconscious attention. The concept of nonconscious attention needs a little explanation because at first look it appears counterintuitive: how can we orient and focus attention on a task without being consciously aware of it? The answer is easy. Our brain does it for us. Imagine there is a bright flash of light, or a sudden loud noise. We immediately attend to those stimuli by turning our heads toward the source. Our attention in this situation is directed to the stimulus without conscious knowledge and effort. However, we eventually become aware of the source, but awareness comes after attention is oriented. There are many other situations that activate the attentional system without our conscious knowledge.

How exactly attention is directed nonconsciously to a task is unclear. The first step in the process, however, is orientation of attention to a target, which can be a sensory input, thought, recalled memory, or future plan. The orienting process is different in overt and in covert attention. Overt orientation involves movement of the eyes in the direction of an object while covert orientation does not involve eye movement.[5] In overt attention, orientation is usually a voluntary process but could also be elicited without conscious awareness. For example, as discussed earlier, patients with blindsight move their eyes in the direction of a source of light even though they are not consciously aware of the source.[6] Since eye movement is considered a defining feature of overt orientation, it could be divided into subtypes: conscious and nonconscious or voluntary and involuntary overt attention. Most of the time we orient attention to a stimulus voluntarily but there are instances when we do it involuntarily by reflex as mentioned above. Both of these are examples of overt orientation, even though reflex is not exactly a conscious action. As mentioned in Chapter 4, Hemineglect, people who have hemineglect do not consciously "see" objects on the neglected field of vision but those objects are attended to and cognitively processed for their meaning.[7] That is why they show a semantic priming effect to stimuli presented in neglected field. As described earlier, when a hemineglect volunteer was shown a good house on the healthy side and a house on fire on the neglected side, she claimed that both houses were identical but did not want to live in the house on fire.[8] It suggests that the house on the neglected side was attended and cognitively processed without conscious awareness. In our daily life we frequently come across similar situations. Even when our attention is focused on a task an unusual stimulus (e.g., a flash of light, or loud noise) makes us orient attention toward that stimulus. The attentional system therefore continuously monitors all stimuli, attended or unattended. That is why our attention is oriented toward an unattended stimulus if it is unusual. This orientation is nonconscious, but if it is considered important, it is brought to conscious awareness possibly by activating the V3A frontal loop described earlier in Chapter 2, Nonconscious memory.[9] If this loop is not activated, we consciously remain unaware of unattended stimuli even though they are monitored by the attentional system.

The idea of the attentional system simultaneously monitoring multiple stimuli appears to contradict classical thinking, which suggests attention can be focused only on one object or task at a time.[10] Let me clarify this apparent contradiction. The classical concept is based on the observation that task performance of volunteers is impaired if their attention is divided. This observation was made in a number of laboratories, including ours. In one of our experiments, the response time of volunteers increased and accuracy decreased when they were asked to perform two tasks simultaneously, as compared to a single task.[11] Increased response time while processing multiple tasks suggests that attention has limited capacity and processes one stimulus at a time. But there is a catch. Even though increased processing time was demonstrated in many well-controlled studies,[12,13] all of these studies were conducted using conscious cognitive tasks and attention was oriented consciously to either one or multiple tasks. The capacity limitation therefore applies to conscious attention processes. It may not necessarily apply to nonconscious attention or processes.

Since multiple tasks that the attentional system monitors remain nonconscious, there may not be a capacity limitation. This possibility is consistent with the observation of an experiment reported by Dijksterhuis described in an earlier chapter.[14] In this experiment volunteers were asked to study either four or 12 attributes of four imaginary cars and rank them based on overall superiority. Ranking was done either immediately or after a delay of four minutes. In the immediate response condition, a decision was made mostly nonconsciously but when they were allowed four minutes to respond, volunteers made a conscious decision with activation of the attentional system. Responses were more accurate in the delayed recall condition. However, when 12 attributes of each car were presented volunteers made a better decision in the immediate recall condition. This experiment demonstrates that in conscious processes, the attentional system can handle only limited pieces of information. However, nonconscious processing remains unaffected by the amount of information. It probably has no capacity imitation. However, note that the findings of Dijksterhuis' experiment need to be interpreted cautiously because other laboratories have failed to replicate some of his results.[15]

Because of the limited capacity, the conscious attention system has a number of constraints. Besides the small amount of information it can handle at a time, it has a limited attention span. That is why it is difficult to remember a number, if it has too many digits. Most of us can remember only four to six digits depending on our age.[16,17] The limit is even smaller if the numbers need to be repeated backward (Fig. 5.2).

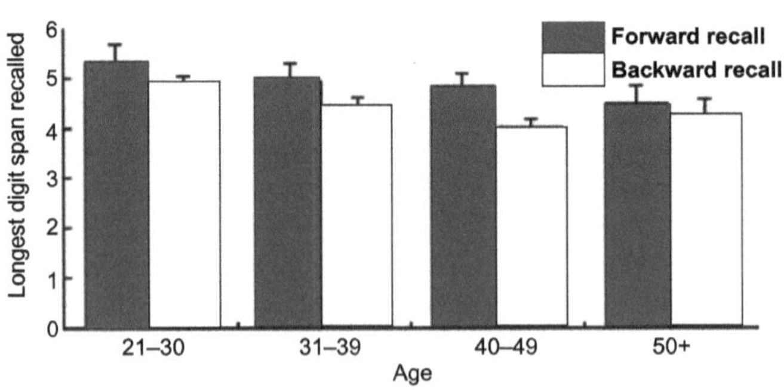

FIGURE 5.2 Digit-span by age. *Modified from a figure by Michaelchilliard via Wikimedia Commons.*

The conscious attention system has temporal limitations too. If attention is focused on a task the system takes about 500 ms to process.[18] Thus, if a volunteer has to respond to a stimulus presented along with a second stimulus, the second stimulus cannot be attended to, if it is presented within the 500 ms refractory period.[19] This delay is primarily because of the time the attentional system takes to orient attention to a new stimulus or to shift attention from one stimulus to the other. If attention is preoriented to a new stimulus the refractory period decreases. Thus, in an experiment a delay in orienting attention could be reduced and even abolished if the location of stimulus is cued before its presentation.[5] Studies suggest that shift of attention does not take 500 ms. The delay is possibly due to activation of the inhibitory system, which does not allow processing of a second stimulus until the previous stimulus has been processed. This ensures that a stimulus does not remain unprocessed or incompletely processed to avoid inaccurate perception.

Even though it is generally believed that the refractory period is around 500 ms,[19] the experiments described in Chapter 2, Nonconscious memory, suggests that it is more likely be around 400 ms, the time a stimulus stays in the V3A frontal loop for attentional processing leading to its conscious appreciation.[9,20–26] This timeframe is consistent with our earlier suggestion. After an extensive review of available data, we suggested that the attentional network does not process a second stimulus if it is presented within 300 ms of the first.[21] It may be the absolute minimum time the inhibitory system needs to ensure full processing of a stimulus. But this timeframe probably varies significantly depending on several factors like illumination, type of stimulus, mental status, etc.

The refractory period therefore is controlled by the inhibitory system, which is primarily a nonconscious process, with or without limited voluntary control. Based on the level of voluntary control there are two kinds of inhibition: executive and automatic. Executive inhibition gives us limited voluntary control but automatic inhibition cannot be controlled voluntarily.

An example of the executive inhibition is a task originally developed in 1974 by American psychologists Barbara and Charles Eriksen at the University of Illinois.[22] In a popular modification of the task called Eriksen's flanker task, volunteers are shown a series of arrowheads pointing either to the right or left. They are asked to indicate the direction of the arrowhead at the center (target arrowhead) using a keypad. The task has a congruent and an incongruent condition. In the congruent condition, all arrowheads point in the same direction (e.g., << << << < or >> >> >> >) while in the incongruent condition the target arrowhead points in a direction different from the other (flanker) arrowheads (e.g., << < > << < or >> > < >> >). Volunteers are asked to respond as quickly and as accurately as possible. The incongruent condition requires them to ignore the nontarget flanker arrowheads and inhibit the response indicated by those arrowheads. In the congruent condition inhibition is not required because all the arrowheads point in the same direction. The response time is expectedly longer in the incongruent condition because it involves the inhibitory system. The response duration indicates whether the inhibitory system is intact or damaged. This task is an example of executive inhibition because the inhibitory delay can be voluntarily altered to an extent by altering the level of attention paid to the task or by practice. Practice-related alteration in duration of executive inhibition has also been studied in another task of executive inhibition the called "go/no-go" task.[23,24] In this task volunteers are shown a series of stimuli and asked to press a key every time they see a specific stimulus. A set of stimuli is designated as no-go and volunteers are supposed to withhold response to those stimuli. The number of false alarms (failure to withhold a response) and response times are recorded. In initial trials the response time is relatively long and volunteers make frequent false alarms. After practice, both response time and frequency of false alarms reduce significantly.

Improvement in these paradigms is primarily due to activity of the basal ganglia. The basal ganglia make templates to respond to different situations. When a task is repeated multiple times, they connect the stimulus and response and make a template that allows execution of a response as soon as the stimulus is presented. It saves time by bypassing several processing steps.

In automatic inhibition there is no voluntary control. Its duration cannot be altered either by manipulating attention or by practice. A task that elicits automatic inhibition was popularized by an American psychologist John Ridley Stroop (Fig. 5.3). For his PhD dissertation Stroop researched the inhibitory system of the nervous system described by physiologists in the late nineteenth century.[25] He found that in 1886 a psychologist James McKean Cattell (Fig. 2.1) discovered that letters are difficult to read if they do not make a word.[26] Stroop extended this discovery at George Peabody College, Nashville, TN in 1935. He found that people also have difficulty naming color of the ink in which a color word (e.g., RED) is printed.[27] The

FIGURE 5.3 John R. Stroop (1897–1973). Credited with popularizing the "Stroop test". *Reproduced from Abilene Christian University, Brown Library Special Collections.*

reason for this difficulty is a conflict between two pieces of information the brain receives: the color of the ink and the meaning of the word. These competing stimuli interfere with the processing, making it difficult to respond. The brain has to inhibit one stimulus (either reading a word or naming the ink color) to respond. It was further observed that inhibiting color naming is relatively easier than inhibiting reading words, probably because we are used to reading words and not naming ink colors. The response time therefore is shorter for reading words than naming colors. This effect is called the word superiority effect or simply the Stroop effect. It is an example of automatic inhibition. Even though the task is named after Stroop it was originally published in German 6 years earlier by E.R. Jaensch.[28] Stroop translated into English.

The Stroop test (Fig. 5.4) is widely used to study the inhibitory system. It works as follows: if the word RED is printed in red ink it is easier to tell the color of the ink as compared to the word RED printed in blue ink. RED in red is a congruent condition in which both word and color are red but RED printed in blue ink is incongruent condition. In this condition the brain gets two different signals: word reads red but ink is blue. The inhibitory system has to suppress the meaning of the word RED in order to name the color. Further, if immediately after this incongruent trial (RED in blue ink) the word BLUE is presented in the red color, volunteers have difficulty naming the color red because in the previous trial the word red was suppressed to

FIGURE 5.4 Stroop test.
Try reading and naming the color of ink in which the words are printed. You will find:
A. It is easier to read a word than to name the color of ink.
B. It is easier to name the color of noncolor words than those of color words.
C. It is easier to name the color if the color and word are the same (RED printed in red ink).
D. It is difficult to name a color if the same color name was suppressed in the previous trial.

Modified from Wikimedia Commons.

name the color of the ink (blue). This difficulty is because the brain had "learned" in the previous trial to suppress the word red. In order to respond in this trial, the system has to "unlearn" (to ignore red) and relearn the color of red ink. It is relatively easy to name color in this trial if the word BLUE is printed in any color other than red. Thus, BLUE printed in green ink would be easier in this example because the inhibitory system had not suppressed the word green and therefore does not need to unlearn and relearn.

Another example of automatic inhibition is a variant of the Stroop effect called negative priming.[29] It was first described in 1985 by a British psychologist named Steven Tipper who was working on his doctoral dissertation at Oxford (he is currently a professor at York University, United Kingdom). Negative priming also demonstrates the effect described above—a stimulus that was ignored in the preceding trial is harder to attend to. In this task, two stimuli—for example, a plus and a minus sign—are presented randomly at one of the four corners of a computer monitor. A volunteer is asked to press a key if a specific stimulus (e.g., plus sign) appears on the monitor. It is difficult for the volunteer to respond to a plus sign on a location if a minus sign was presented at the same location in the previous trial. Since the stimulus at the target location was ignored in the previous trial, the inhibitory system learns to ignore that location as a possible location for the target. In order to respond, it needs to unlearn and relearn that the previously ignored location could be a possible target location. Since this type of inhibition cannot be voluntarily controlled, it is also an example of automatic inhibition.

These examples suggest that the inhibitory system makes the brain recognize a pattern by ignoring insignificant and attending to significant stimuli. It also helps the brain find the most probable location of a target. With each experience the brain learns pattern of events and stimuli. Thus, in negative priming when a specific location was ignored in a trial, the inhibitory system marks the location as insignificant and and uses attention resources to focus on other possible locations.

This effect was further elaborated in a task developed by Michael Posner (Posner cuing task).[5] In this task a volunteer is asked to look at a fixation point (a cross-mark or dot placed at the center of a computer monitor). A stimulus appears either on the right or left of the fixation point. The volunteer is asked to press a key as quickly and as accurately as possible as soon as the stimulus appears. An arrow pointing either to the right or left is presented before the stimulus that appears either in the direction of the arrow (valid trials) or in opposite direction (invalid trials). As expected, the response time is shorter in valid trials. In these trials, the attentional system begins orienting attention as soon as the arrow appears, several milliseconds before appearance of the target. Thus, the time of orientation begins with the appearance of the arrow. In invalid trials volunteers had to reorient attention to the other side, thereby incurring delay. It was further observed[30] that valid trials have a response time advantage only if the arrow is presented between

100 and 300 ms before appearance of the stimulus. If stimulus is presented 500 ms after the arrow, the response time advantage disappears. In this condition, it actually takes longer to orient attention in valid trials compared to invalid trials. This phenomenon is called inhibition of return.[30] Valid trials take longer in this condition because attention is initially oriented toward the cue but because the target does not appear within 300 ms, the focus of attention goes back to the fixation point, expecting the stimulus to appear on the other side. If the stimulus appears at the previously cued location, attention has to disengage from the expected location and reorient toward the cued location, thereby increasing response time.

Interestingly, inhibition of return is observed only when the target stimulus is externally triggered and require the eyes to move. If it does not require eye movement (as in an endogenous target) inhibition of return is not observed.[31] It was therefore suggested that inhibition of return involves oculomotor response[32] and its latency is due to the conscious effort required to move the eyes. If that were the case, nonconscious attention not requiring conscious movement of eyes should not have inhibition of return. Attention can be moved to an intended stimulus nonconsciously without the delay imposed by inhibition of return. Inhibition of return therefore is a limitation of conscious attention but not of nonconscious attention.

Inhibition of return and negative priming helps focus attention in the direction in which the target is most likely to appear. It reduces the time needed to orient attention, which in turn reduces the time needed for processing a stimulus. The absence of negative priming and inhibition of return indicate a damaged inhibitory system. It is damaged in many psychiatric and neuropsychiatric conditions. Most notably, people with schizophrenia do not show either negative priming or inhibition of return. It is one of the reasons they have impaired cognition. It is possible that absence of inhibition of return is due to their inability to disengage attention from the originally engaged location. This compromises their ability to broaden the search for a target and modify the strategy if the target is not at the expected location and the search requires shifting attention to another direction.

The neural network that controls inhibition of return is unclear but it appears that the superior colliculus, parietal lobe, and thalamus are involved in its processing.[33] Damage to the superior colliculus therefore abolishes inhibition of return.[34] It also suggests that the latency of inhibition of return is primarily due to delay in the eye movement. The oculomotor nerve that controls eye movements originates in the vicinity of the superior colliculus.

The superior colliculi play an important role in shifting attention because presence of the nucleus of the oculomotor nerve in the vicinity. Shift of attention however, is initiated at the parietal cortex, as we suggested earlier.[3] That is why parietal lesions cause hemineglect, a condition in which an individual does not pay attention to objects in one half of the field of vision as discussed in Chapter 4, Hemineglect. People with parietal lesions take almost three times longer to disengage and reengage attention to a new target.[35] The parietal cortex thus controls attentional shift. The right parietal appears to specialize in this function because increased activation in this area is observed when attention is shifted in either direction—right or left. The left parietal is activated only when the shift is toward the right side.[36] Because of bilateral control, damage to the right parietal cortex causes hemineglect, even if the left is intact. This rarely happens after left parietal damage.[37] Another reason for the lateralized effect is the difference in attentional processing of the right and left parietal lobes. If a large picture consisting of several small pictures in shown, an individual with left parietal damage would neglect small pictures and report the large picture but those with right parietal damage would report small pictures and neglect the big picture. Thus, the right parietal looks at the forest while the left looks at the trees in the forest.[38] That is why an individual does not see the whole picture following right parietal damage. It limits perceptual ability leading to hemineglect.[37] In addition to the superior colliculus and parietal lobe, the nucleus pulvinar (a part of the thalamus) is also involved in attentional shifts. It is activated when attention is shifted from one location to the other.[39]

Attention shift involves multiple brain areas because it consists of several components, like reorientation of gaze and head, suppression of signals from the unattended location, and amplification of signals from the attended target. When visual attention is shifted from one object to the other, the neurons of visual association area in monkeys stop responding to previously attended stimulus and begin responding to the newly attended object. Similar changes are not observed in early visual neurons of the retina, lateral geniculate body, or primary visual cortex. These neurons respond to both attended and unattended stimuli with the same intensity. Attention processing therefore begins in the visual association area, including area V3A as discussed earlier. From this area signals of attended objects are relayed to the frontal cortex possibly using the V3A frontal loop. The cognitive significance of these signals are then analyzed in the frontal cortex and executive network.[40]

Based on the findings of our experiments, it appears that the executive system processes the stimulus if it is unusual or unexpected. In an experiment conducted in Michael Posner's laboratory using event-related potentials we asked volunteers to generate a verb for a noun and respond after a fixed time delay (1.5 seconds). The task activated the frontal cortex and cingulate gyrus.[41] These areas are also activated while resolving a cognitive conflict. For example, if an ambiguous word like "bank" needs to be interpreted, the executive network defines the context and decides if it refers to a money transaction or a river. The network stops responding if it is used in the same context over and over again because then the task becomes automatic. Similarly, cingulate activation is observed in the Stroop task, which also involves cognitive conflict between reading the word and naming the color of ink.[42,43]

Experiments conducted in our laboratory indicate that the basal ganglia are also involved in attentional modulation and inhibitory control. Basal ganglia probably have more influence on inhibition of attention to the nontarget rather than its facilitation on the target.[44–46] This conclusion is based on the data we acquired in a series of neuroimaging experiments in which we used Eriksen's flanker task (described earlier in this chapter) to study the inhibitory system. In this task, volunteers were required to press a key to indicate the direction of arrowheads presented in quick succession. In the incongruent condition of the task (described above), volunteers get two contradictory signals: the target arrowhead indicates one direction and the flanker arrowheads indicate the other. To make a response they have to inhibit signals coming from the flanker arrowheads and focus on the target arrowhead. In this condition we observed increased activity in the basal ganglia, anterior cingulate, and frontal cortex.[46] Since the flanker task elicits cognitive processing associated with inhibition of irrelevant response, areas activated in the task are involved in suppression of the nontarget instead of facilitation of target signals.

The increased activity of the basal ganglia observed in this experiment suggested that dopamine may be involved in processing of inhibition because these ganglia have high density of dopamine receptor. To confirm the suggestion, we studied dopamine neurotransmission during inhibition using a novel neuroimaging technique we developed to detect, map, and measure acute changes in dopamine concentration in live human brain. The technique is called single-scan dynamic molecular imaging technique (SDMIT) or simply neurotransmitter imaging. It has been extensively used by us[47–58] and others[59,60] to detect dopamine released acutely in the brain. In this experiment after volunteers were positioned in positron emission tomography (PET) camera, they received an intravenous injection of a dopamine receptor ligand raclopride, which binds to dopamine receptors in the brain. The ligand was radiolabeled with ^{11}C to allow monitoring of its concentration by a PET camera. After the injection, volunteers were asked to perform Eriksen's flanker task. During task performance ligand concentration was measured dynamically. In the first 20 minutes they performed the task under the congruent condition and thereafter under the incongruent condition, which activates the inhibitory system to inhibit nonrelevant response as discussed above. Analysis of the data indicated decreased ligand concentration in the left caudate and in the putamen of both hemispheres in the incongruent condition (Fig. 5.5), indicating dopamine release in these areas. The results provided direct evidence of involvement of dopamine neurotransmission in inhibition of nonrelevant stimuli.[44,45]

Dopaminergic control of attention has been under controversy for quite some time because results of the experiments conducted to examine its activity in people who have difficulty focusing and sustaining attention were inconclusive. Thus, individuals with attention deficit hyperactivity disorder

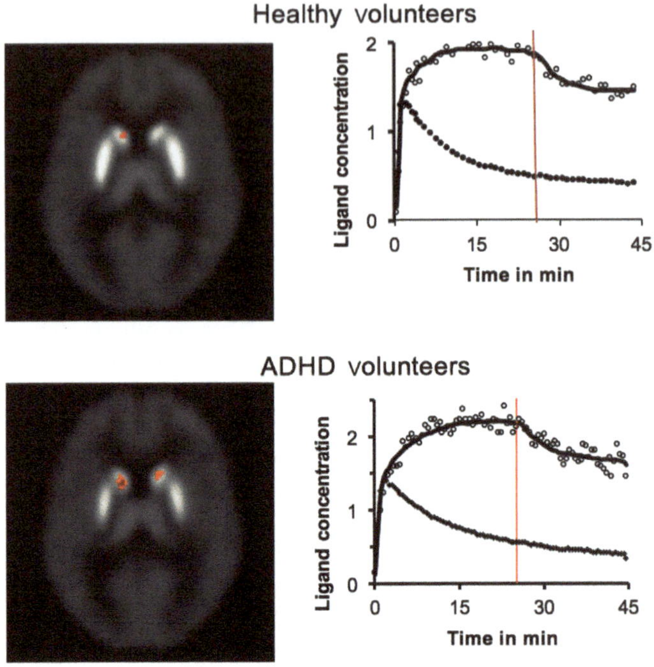

FIGURE 5.5 Dopamine release during performance of Eriksen's flanker task in healthy volunteers and in individuals with attention deficit hyperactivity disorder (ADHD). Time activity curves: When the task condition was changed from congruent to incongruent (*red line*), the ligand (^{11}C-raclopride) concentration reduced significantly in the left caudate in healthy volunteers and bilaterally in the cuadate of individuals with ADHD. *Adapted from Badgaiyan RD, Sinha S, Sajjad M, Wack DS. Attenuated tonic and enhanced phasic release of dopamine in attention deficit hyperactivity disorder. PLoS One 2015;**10**(9), e0137326.*

(ADHD) have impaired ability to focus attention on a target and have difficulty ignoring nonrelevant stimuli. Studies on these individuals have reported data suggesting either increased or decreased dopamine.[61] These experiments reported contradictory data because they used indirect methods to measure dopamine in the brain. They had to use the indirect method because a direct method to measure dopamine in the live human brain was unavailable until we developed the neurotransmitter imaging technique SDMIT described above.[57,58] The indirect methods used by earlier investigators probably detected different aspects of dopaminergic activity. Dopamine is released in the brain in two stages: one at rest (tonic release) and the other during performance of a dopamine-dependent task (phasic release). Since there is a reciprocal relationship between tonic and phasic release,[62] if tonic release is attenuated, the phasic will be enhanced. Indirect methods would therefore come to contradictory conclusions depending on detection of either tonic or phasic release. Since most of these methods cannot differentiate between tonic and phasic release, it is difficult to know the status of dopamine in ADHD using those data.

However, tonic and phasic release of dopamine can be separately detected, mapped, and measured using SDMIT. We therefore used this technique to study dopamine neurotransmission in ADHD volunteers at rest (tonic release) and during performance of a dopamine-dependent Eriksen's flanker task (phasic release).[44,45] As mentioned above, in the incongruent condition of the task dopamine was released in the left caudate and in the putamen of healthy volunteers (Fig. 5.5). In ADHD volunteers, it was released in the caudate of both hemispheres and in the putamen (Fig. 5.5), indicating additional dopamine release in the right caudate (Figs. 5.6 and 5.7). Further, as compared to healthy volunteers there was significant attenuation of tonic dopamine release in the right caudate (Fig. 5.6). ADHD individuals therefore have low levels of dopamine at rest in the right caudate. To compensate, an increased amount is released during task performance. The low level of tonic dopamine in the right caudate therefore makes inhibition difficult in people with ADHD.

In this study we obtained direct evidence of involvement of dopamine in attentional processing and helped address the controversy concerning the nature of dysregulated dopamine neurotransmission in people with ADHD. The results also suggest that attenuated tonic dopamine in the right caudate may be the target for therapeutic intervention in individuals with ADHD. The intervention should be directed at enhancing the tonic level. It could be accomplished by limiting dopamine reuptake into presynaptic neurons. Normally, most of the dopamine released in the synapse from presynaptic neurons is transported back to those neurons by a protein called dopamine transporter (DAT) (Fig. 5.8). These transporters bind to dopamine molecules

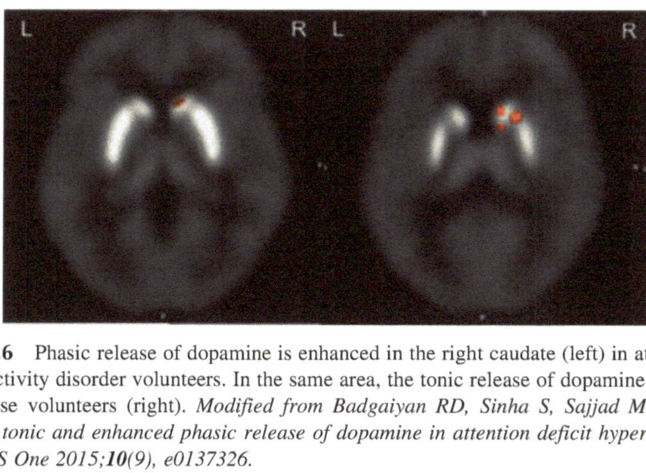

FIGURE 5.6 Phasic release of dopamine is enhanced in the right caudate (left) in attention deficit hyperactivity disorder volunteers. In the same area, the tonic release of dopamine was attenuated in these volunteers (right). *Modified from Badgaiyan RD, Sinha S, Sajjad M, Wack DS. Attenuated tonic and enhanced phasic release of dopamine in attention deficit hyperactivity disorder. PLoS One 2015;**10**(9), e0137326.*

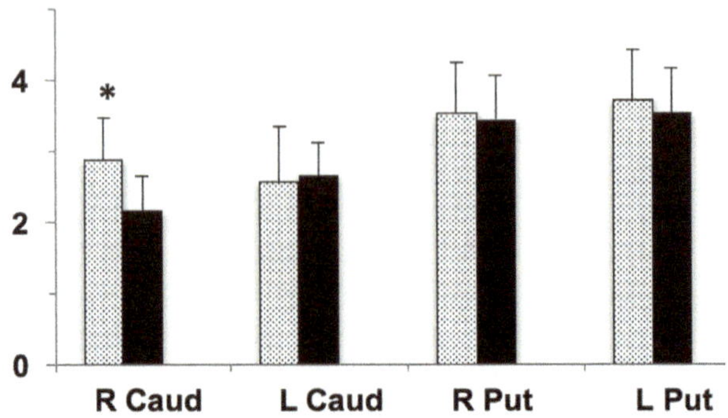

FIGURE 5.7 Ligand binding in healthy (*light bars*) and attention deficit hyperactivity disorder (*dark bars*) volunteers in the right (R) and left (L) caudate (Caud) and putamen (Put). The difference was significant ($^{*}P < .004$) only in the right caudate. Lower binding indicates higher dopamine release. *Reproduced from Badgaiyan RD, Sinha S, Sajjad M, Wack DS. Attenuated tonic and enhanced phasic release of dopamine in attention deficit hyperactivity disorder. PLoS One 2015;**10**(9), e0137326.*

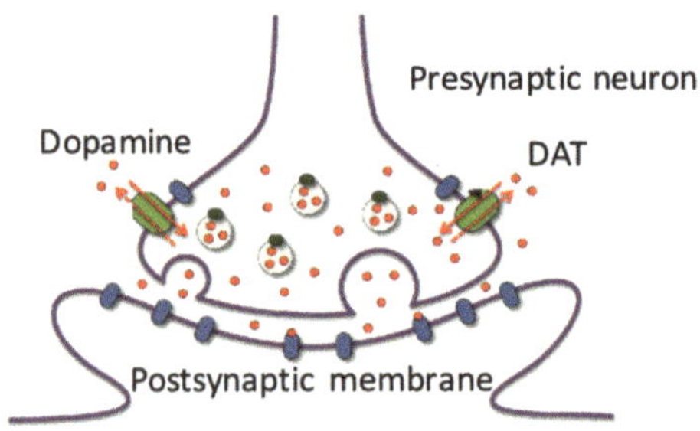

FIGURE 5.8 Dopamine transporters (DAT) transfer dopamine (*red dots*) released in the synapse back to presynaptic neurons. Blue circles represent dopamine receptors.

and attach to DAT receptors located on the wall of presynaptic neurons. If DAT receptors are blocked pharmacologically, reuptake of dopamine is also blocked, resulting in its reduced reuptake. Reduction in reuptake increases dopamine level in the synaptic cleft increasing its tonic pool. Therefore, agents that block DAT receptors are most effective clinically in ADHD patients. That is why the efficacy of current ADHD medications depends on their ability to block DAT. It is therefore not a coincidence that the most potent DAT blocker, methylphenidate (Ritalin), is also the most effective clinically.[63] The results also explain why dopaminergic agents like L-dopa that do not block DAT are not clinically effective in treating ADHD.[64]

Thus, the attentional system can be controlled by manipulating the inhibitory system. Besides controlling the attentional system, the inhibitory system makes sure that a stimulus is fully processed before the next one is received. It also prevents information overload by temporally spacing stimuli. However, the most important function of the system is probably the pattern recognition. By making the brain learn to ignore insignificant and attend to significant stimuli, it helps find the most probable location of a target.

Since attention significantly influences cognition and behavior, it is important to have a better understanding of attentional control. We can then exploit its potential to enhance our mental and physical abilities.

Bibliography

1. James W. *Principles of psychology*. Holt; 1890.
2. Patañjali, Dvivedi MN, Subrahmanya Sastri S. *The yoga-sūtras of patañjali*. Theosophical Publishing House; 1934.
3. Posner M, Badgaiyan R. Attention and neural networks. In: Parks RW, Levine DS, editors. *Fundamentals of neural network modeling: neuropsychology and cognitive neuroscience*. The MIT Press; 1998. p. 61–76.
4. Badgaiyan RD. Nonconscious perception, conscious awareness and attention. *Conscious Cogn* 2012;**21**(1):584–6.
5. Posner MI. Orienting of attention. *Quart J Exp Psychol* 1980;**32**:3–25.
6. Poppel E, Held R, Frost D. Residual visual function after brain wounds involving the central visual pathways in man. *Nature* 1973;**243**(405):295–6.
7. McGlinchey-Berroth R, Milberg WP, Verfaellie M, Alexander M, Kilduff P. Semantic priming in the neglected field: evidence from a lexical decision task. *Cogn Neuropsychol* 1993;**10**:79–108.
8. Marshall JC, Halligan PW. Blindsight and insight in visuo-spatial neglect. *Nature* 1988;**336** (6201):766–7.
9. Badgaiyan RD. Conscious awareness of retrieval: an exploration of the cortical connectivity. *Int J Psychophysiol* 2005;**55**(2):257–62.
10. Posner MI. Structure and functions of selective attention. In: Boll T, Bryant B, editors. *The master lecture series, clinical neuropsychology and brain function: research, measurement, and practice*, vol. 7. Americal Psychological Association; 1988. p. 173–202.
11. Badgaiyan RD, Schacter DL, Alpert NM. Priming within and across modalities: exploring the nature of rCBF increases and decreases. *Neuroimage* 2001;**13**(2):272–82.
12. Posner MI, Petersen SE. The attention system of the human brain. *Annu Rev Neurosci* 1990;**13**:25–42.
13. Posner MI. Attention before and during the decade of the brain. In: Mangun GR, Hillyard SA, Luck SJ, editors. *Attention and performance*. MIT Press; 1993. p. 343–50.
14. Dijksterhuis A, Bos MW, Nordgren LF, van Baaren RB. On making the right choice: the deliberation-without-attention effect. *Science* 2006;**311**(5763):1005–7.
15. Shanks DR, Newell BR, Lee EH, Balakrishnan D, Ekelund L, Cenac Z, et al. Priming intelligent behavior: an elusive phenomenon. *PLoS One* 2013;**8**(4):e56515.
16. Knight RT. Evoked potential studies of attention capacity in human frontal lobe lesions. In: Eisenberg HM, Benton AL, Levin HS, editors. *Frontal lobe function and dysfunction*. New York, NY: Oxford University Press; 1991. p. 139–53.
17. Jonkman L, Kemner C, Verbaten M, Van Engeland H, Camfferman G, Buitelaar J, et al. Attentional capacity, a probe erp study: differences between children with attention-deficit hyperactivity disorder and normal control children and effects of methylphenidate. *Psychophysiology* 2000;**37**(3):334–46.

18. Posner MI, Snyder CRR. Attentional and cognitive control. In: Solso R, editor. *Cogntive psychology*. Lawrence Erlbaum Associates; 1975. p. 55–85.
19. Dux PE, Marois R. The attentional blink: a review of data and theory. *Atten Percept Psychophys* 2009;**71**(8):1683–700.
20. Badgaiyan RD, Posner MI. Time course of cortical activations in implicit and explicit recall. *J Neurosci* 1997;**17**(12):4904–13.
21. Badgaiyan RD. Nonconscious processing and a novel target for schizophrenia research. *Open J Psychiatry* 2012;**2**(4A).
22. Eriksen BA, Eriksen CW. Effect of noise letters upon the identification of a target letter in a non search task. *Percep Psychophys* 1974;**16**:143–9.
23. Pérez-Fernández C, Cánovas R, Moreno M, Sánchez-Santed F, Flores P. Go/nogo training improves executive functions in an 8-year-old child born preterm. *Revista de Psicología Clínica con Niños y Adolescentes* 2017;**4**(3):60–6.
24. Rueda MR, Checa P, Combita LM. Enhanced efficiency of the executive attention network after training in preschool children: immediate changes and effects after two months. *Dev Cogn Neurosci* 2012;**2**(Suppl. 1):S192–204.
25. Uchizono KJ. *Excitation and inhibition: synaptic morphology*. Igaku Shoin; 1975.
26. Cattell JM. The time it takes to see and name objects. *Mind* 1886;**os-XI**(41):63–5.
27. Stroop JR. Studies of interference in serial verbal reactions. *J Exp Psychol* 1935;**18**:643–62.
28. Jaensch ER. *Grundformen menschlichen seins (mit berücksichtigung ihrer beziehungen zu biologie und medizin, zu kulturphilosophie und pädagogik)*. O. Elsner Verlagsgesellschaft MBH; 1929.
29. Tipper SP. The negative priming effect: inhibitory effects of ignored primes. *Quart J Exp Psychol* 1985;**37A**:571–90.
30. Posner MI, Cohen Y. Components of visual orienting. In: Bouma H, Bouwhuis D, editors. *Attention and performance X: control of language processes*. Erlbaum; 1984. p. 531–56.
31. Maylor EA, Hockey R. Inhibitory component of externally controlled covert orienting in visual space. *J Exp Psychol Human Percept Perf* 1985;**11**(6):777–87.
32. Rafal RD, Calabresi PA, Brennan CW, Sciolto TK. Saccade preparation inhibits reorienting to recently attended locations. *J Exp Psychol Human Percept Perf* 1989;**15**(4):673–85.
33. Posner MI, Rafal RD, Choate L, Vaughn J. Inhibition of return: neural basis and function. *Cogn Neuropsychol* 1995;**2**:211–28.
34. Sapir A, Soroker N, Berger A, Henik A. Inhibition of return in spatial attention: direct evidence for collicular generation. *Nat Neurosci* 1999;**2**(12):1053–4.
35. Posner MI, Early TS, Reiman E, Pardo PJ, Dhawan M. Asymmetries in hemispheric control of attention in schizophrenia. *Arch Gen Psychiatry* 1988;**45**:814–21.
36. Corbetta M, Miezin FM, Shulman GL, Petersen SE. A PET study of visuospatial attention. *J Neurosci* 1993;**13**(3):1202–26.
37. Vallar G. Spatial hemineglect in humans. *Trends Cogn Sci* 1998;**2**(3):87–97.
38. Doricchi F, Incoccia C. Seeing only the right half of the forest but cutting down all the trees? *Nature* 1998;**394**(6688):75–8.
39. Grieve KL, Acuna C, Cudeiro J. The primate pulvinar nuclei: vision and action. *Trends Neurosci* 2000;**23**(1):35–9.
40. Badgaiyan RD. Executive control, willed actions, and nonconscious processing. *Hum Brain Mapping* 2000;**9**(1):38–41.
41. Badgaiyan RD, Posner MI. Mapping the cingulate cortex in response selection and monitoring. *Neuroimage* 1998;**7**(3):255–60.

42. Pardo JV, Pardo PJ, Janer KW, Raichle ME. The anterior cingulate cortex mediates processing selection in the stroop attentional conflict paradigm. *Proc Natl Proc Natl Acad Sci USA* 1990;**87**(1):256–9.
43. MacDonald III AW, Cohen JD, Stenger VA, Carter CS. Dissociating the role of the dorsolateral prefrontal and anterior cingulate cortex in cognitive control. *Science* 2000;**288** (5472):1835–8.
44. Badgaiyan RD, Sinha S, Sajjad M, Wack DS. Attenuated tonic and enhanced phasic release of dopamine in attention deficit hyperactivity disorder. *PLoS One* 2015;**10**(9): e0137326.
45. Badgaiyan RD, Wack D. Evidence of dopaminergic processing of executive inhibition. *PLoS One* 2011;**6**(12):e28075.
46. Casey BJ, Thomas KM, Welsh TF, Badgaiyan RD, Eccard CH, Jennings JR, et al. Dissociation of response conflict, attentional selection, and expectancy with functional magnetic resonance imaging. *Proc Natl Acad Sci USA* 2000;**97**(15):8728–33.
47. Badgaiyan RD. Imaging dopamine neurotransmission in live human brain. *Prog Brain Res* 2014;**211**:165–82.
48. Badgaiyan RD. Detection of dopamine neurotransmission in "real time.". *Front Neurosci* 2013;**7**:125.
49. Badgaiyan RD. Neurotransmitter imaging: current status and challenges. *Curr Med Imaging Rev* 2011;**7**:96–8.
50. Badgaiyan RD. Neurotransmitter imaging: basic concepts and future perspectives. *Curr Med Imaging Rev* 2011;**7**:98–103.
51. Badgaiyan RD. Dopamine is released in the striatum during human emotional processing. *Neuroreport* 2010;**21**:1172–6.
52. Badgaiyan RD, Fischman AJ, Alpert NM. Dopamine release during human emotional processing. *Neuroimage* 2009;**47**(4):2041–5.
53. Badgaiyan RD, Fischman AJ, Alpert NM. Explicit motor memory activates the striatal dopamine system. *Neuroreport* 2008;**19**(4):409–12.
54. Badgaiyan RD, Fischman AJ, Alpert NM. Striatal dopamine release in sequential learning. *Neuroimage* 2007;**38**(3):549–56.
55. Fischman AJ, Badgaiyan RD. Neurotransmitter imaging. In: Charron M, editor. *Pediatric PET.* Springer; 2006. p. 385–403.
56. Badgaiyan RD, Fischman AJ, Alpert NM. Detection of striatal dopamine released during an explicit motor memory task. *J Nucl Med* 2005;**46**(Suppl. 2):213.
57. Badgaiyan RD, Fischman AJ, Alpert NM. Striatal dopamine release during unrewarded motor task in human volunteers. *Neuroreport* 2003;**14**(11):1421–4.
58. Alpert NM, Badgaiyan RD, Livini E, Fischman AJ. A novel method for noninvasive detection of neuromodulatory changes in specific neurotransmitter systems. *Neuroimage* 2003;**19**(3):1049–60.
59. Christian B, Lehrer D, Shi B, Narayanan T, Strohmeyer P, Buchsbaum M, et al. Measuring dopamine neuromodulation in the thalamus: using [f-18]fallypride PET to study dopamine release during a spatial attention task. *Neuroimage* 2006;**31**(1):139–52.
60. Backman L, Nyberg L, Soveri A, Johansson J, Andersson M, Dahlin E, et al. Effects of working-memory training on striatal dopamine release. *Science* 2011;**333** (6043):718.
61. Genro JP, Kieling C, Rohde LA, Hutz MH. Attention-deficit/hyperactivity disorder and the dopaminergic hypotheses. *Expert Rev Neurother* 2010;**10**(4):587–601.

62. Grace AA. Phasic versus tonic dopamine release and the modulation of dopamine system responsivity: a hypothesis for the etiology of schizophrenia. *Neuroscience* 1991;**41**(1):1–24.
63. Volkow N, Wang G, Fowler J, Ding Y. Imaging the effects of methylphenidate on brain dopamine: new model on its therapeutic actions for attention-deficit/hyperactivity disorder. *Biol Psychiatry* 2005;**57**(11):1410–15.
64. Langer D, Rapoport J, Brown G, Ebert M, Bunney WJ. Behavioral effects of carbidopa/levodopa in hyperactive boys. *J Am Acad Child Psych* 1982;**21**(1):10–18.

Chapter 6

Decision making

Every day we make dozens, probably hundreds, of decisions. These decisions include not only major decisions like selecting a partner or choosing a job, but also small and mundane decisions like what to eat for breakfast, where to sit, who to talk to, where to go for a walk, etc. It seems we make those decisions with full conscious awareness based on the available options. But believe it or not, most of those decisions are made nonconsciously and conveyed to the conscious mind that then 'owns' them. That is the reason we wonder at times why we made a decision in certain way even though we made them with full conscious awareness.

The influence of nonconscious mind over conscious decisions is best illustrated by our biases and preferences. All of us have preferences and aversions to certain foods, objects, people, colors, places, etc. We like to eat certain types of food one day and other type the next day. We decide not to eat certain foods at all, and make other foods our favorites. But how do we decide not to "like" certain foods, colors, or types of music, for example? Obviously, for the most part we do not have a good reason or justification. Sometimes we "make up" a reason to convince ourselves that we are making a rationale, informed and conscious decision. For example, "I like the color blue because it is soothing" but other colors are equally or probably more soothing. Why only blue? We don't know the real reason for preferences because those decisions are made nonconsciously. However, the nonconscious mind may have a "reason" why we should like or dislike something, but that reason never enters our conscious mind.

We do not know how the nonconscious mind decides which food we should like and which one to dislike. It is probably based on past experience. As an example, let us assume I do not like potatoes; if someone asks why I do not like them, perhaps the only answer I would have is they do not taste good. But that is not the real answer because I do not know why they do not taste good. My nonconscious mind probably "knows" the answer. It might have retained a memory of a past event that I may have forgotten. It is also possible that sometime in the past I ate potatoes and something bad happened immediately after. The nonconscious mind makes an association between the two. Even though I do not consciously remember the event or make a connection, the nonconscious mind remembers it and makes me

Neuroscience of the Nonconscious Mind. DOI: https://doi.org/10.1016/B978-0-12-816115-9.00006-1

dislike potatoes to avoid the event that followed in the past. The same could be true for our other preferences. Thus, favorable events following eating certain foods or wearing clothes of a certain color may nonconsciously make us like that food or color. Therefore, most decisions to eat certain foods or buy certain clothes of a particular color or design are made nonconsciously but executed consciously and regarded as conscious decisions.

The role of the nonconscious mind in the decision-making process has been studied in laboratory settings by many investigators. In one of those experiments conducted by Dijksterhuis described earlier[1] volunteers were asked to determine the superiority of cars based on either fouror 12 attributes. They were also asked to respond immediately and after deliberation. In the immediate response condition the decision was made nonconsciously because it did not allow activation of the attentional network needed for conscious processing as discussed earlier.[2] Decisions made after deliberation allowed the volunteers to make conscious decisions. When four attributes were given, participants made better decisions after deliberation but when 12 attributes were presented, 70% of volunteers made better decisions in the immediate response condition. Itsuggests that the nonconscious approach is superior when the number of variables is large. This finding is significant because most real-life decisions involve a large number of variables. For example, to decide which job to accept, variables that need to be considered could include the nature of the job, boss, future prospects, place of work, effect of the job on family and friends, and so on. The results of Dijksterhuis' experiment are convincing but needs cautious interpretation because other laboratories have failed to replicate many of his findings.[3]

In addition to being more accurate under certain situations, the nonconscious mind makes decisions much quicker than the conscious mind. This quality of the nonconscious mind was demonstrated in an experiment designed by Antoine Bechara, an American psychologist who developed a gambling task while working at the University of Iowa.[4] In this task, called the Iowa gambling task, four decks of cards are prepared. Each card indicates the amount of money a volunteer will gain or lose by picking it. Two of the decks are labeled as "good" decks. Cards of these decks pay more money than they take, if chosen repeatedly. The other two decks are labeled "bad." The cards in the good decks give money in small amounts and take away small amounts, but the cards in the bad decks give a large amount of money but take away even more. Volunteers are asked to pick cards from the deck of their choice and to win as much money as possible. The experiment is terminated when a volunteer has picked 100 cards. After the first 20 and then every 10 picks, Bechara interrupted the experiment and asked volunteers if they know which decks were good and which were bad. Their skin conductance (also called the galvanic skin response, or GSR), which indicates anxiety, was measured continuously during the experiment. Skin conductance measures resistance to flow of current between two points on the skin. It is high when the skin is dry and low when it is moist. Under high anxiety and emotional situations the skin sweats (to cool off the body and prepare for the action), reducing the resistance. In Bechara's experiment the skin conductance started showing "anticipatory response" (low GSR) just before picking cards from the bad decks after volunteers had picked only 10 cards. At that point, they had no conscious knowledge of good or bad decks.[5] After picking 50 cards volunteers started having a "hunch" that two of the decks were better than the other two and after 80 cards most of them were able to identify the good and bad decks. The GSR changes observed after 10 picks suggest that the nonconscious mind "knew" which decks were good and which were bad at a very early stage. But this knowledge was not made available to the conscious mind. Therefore, volunteers had no conscious knowledge at that point. They gained that knowledge consciously much later.

Another finding of this experiment concerns conscious choice. After 80 picks almost all volunteers started picking cards from good decks even though 3 out of 10 volunteers did not know which decks were good. Picking cards from good decks without conscious knowledge indicates that their conscious decision to choose cards was made by the nonconscious mind. Further, in this experiment Bechara observed an interesting pattern in people with damaged ventral frontal cortex.[6] These individuals did not show an anticipatory skin conductance response and did not consciously know which decks were good, with the exception of three people that were able to identify good and bad decks. However, despite having conscious knowledge, they kept picking cards from bad decks and ended up losing money. This behavior illustrates the dissociation between the processes that control nonconscious knowledge, conscious knowledge, decisions, and actions.

Observations of this experiment suggest that the brain makes decision at a nonconscious level at a very early stage of the decision-making process. Most of the time we make the same decision consciously much later. Even though we are not aware of a nonconscious decision, it probably helps us build confidence in our decision. We gain confidence if our conscious decision is consistent with the nonconscious one. If the nonconscious mind has made a different decision or it has not made a decision at all, our level of confidence remains low. This is the reason why people with lesions in the ventral frontal cortex keep picking cards from bad decks even though they consciously know which decks are good. Since these individuals were unable to make nonconscious decisions (indicated by the lack of skin conductance response), their confidence in conscious decision was not high enough to make them act on it.

This experiment underscores the interaction between conscious and nonconscious decisions. To understand the relationship between the two, we need to discuss how these decisions are made. It appears that the conscious decision-making process begins with nonconscious retrieval of memories of relevant past events, actions and outcomes. This information is probably retained as modules. Each experience or event creates a module and when a similar event is encountered, the module is activated and the appropriate action included in it is activated, allowing us to make quick decisions. However, this decision-making process remains nonconscious. In order for us to make the decision available to conscious awareness, it has to go through a second level of processing, which probably involves the V3A frontal loop and attentional resources. During this process, the conscious mind modifies the decision to make it consistent with cultural norms and also to suit one's mood, beliefs, and a host of other personal and moral factors. If modification is not significant both conscious and nonconscious decisions remain in agreement. This agreement provides confidence in the decision. If conscious processes make substantial modification and the decision is not consistent with the nonconscious decision, we have low confidence.

It is unclear to what extent the nonconscious mind influences our conscious decisions. It probably varies based on the nature of decision. If a decision has a significant emotional component, the nonconscious processes probably play a greater role. Decisions involving a large number of variables may also use nonconscious processes because of their ability to use multiple variables simultaneously. Additionally, since past experience is an important input in the decision-making process, nonconscious processes have an advantage because of their ability to access a larger pool of information. The nonconscious mind can access conscious memory and also memories that have been forgotten or were never consciously registered. It thus has a greater amount of information than the conscious mind.

This discussion does not imply that nonconscious decisions are always superior to conscious decisions. Under certain conditions conscious decisions are better. If the number of variables needed to make a decision is limited and all variables are known, conscious decisions are superior as demonstrated in Dijksterhuis's experiment described above.[1] The findings of this experiment are consistent with the observation of another Dutch psychologist Silvia Mamede[7] who asked expert and novice doctors to diagnose a condition based on given clinical data. They were asked to make a diagnosis either immediately or after deliberation. As mentioned above, decisions made immediately depend on nonconscious processes while those made after deliberation involve conscious processes. She found that expert doctors made 50% better diagnoses after deliberation as compared to diagnoses made without deliberation. Conscious decisions were therefore superior to nonconscious ones in this situation. However, the accuracy of novice doctors' diagnoses did not improve after deliberation. They did not benefit from conscious decision-making processes because they had limited pools of knowledge for making decisions. Thus, conscious decisions are better if all pieces of information are available. The availability of information therefore is an important determinant of the quality of conscious decisions.

Under ideal conditions, conscious decisions are reliable but if conditions are not ideal, if all pieces of information are not available, or if the attentional system is not adequately activated, the conscious mind cannot make good decisions. As discussed in previous chapters, activation of the attentional system is required for conscious processes but not for nonconscious processes.[2,8] We also discussed previously that the attentional system cannot process multiple pieces of information simultaneously. It therefore limits the accuracy of conscious decision, particularly if the decision has to be made quickly and involves multiple inputs. John Payne and James Bettman of Duke University suggested that conscious decisions improve if unlimited time is available. They imply that even when dealing with a large number of variables, conscious decisions may not be inferior to nonconscious ones if enough time and relevant information is available.[9] But it is not clear how much time is enough.

Timing appears to have an inverted U-shaped relation with the quality of decision. The best decisions are made if the time is "optimal." If it is too little or too much the decision may not be the best. Most of us are familiar with the idea that excessive pondering leads to focusing on minor issues and often leads to bad decisions.[10] This was experimentally demonstrated by Wilson and Schooler.[11] In an experiment they asked college students to rate different brands of strawberry jam. One group was only had to give their ratings while the other group had to list reasons for the rating. Interestingly, ratings of the group that was not asked to think about a reason were closer to the ratings of taste analysts. The group that was not asked to give reason made nonconscious decision but students who had to give a reason were forced to make the rating based on conscious decision, after pondering and focusing on different aspects.

Apart from uncertainty about optimal time, limited capacity of the attentional system discussed in the previous chapter also affects conscious decisions.[12] If a decision requires analysis of multiple pieces of information, the attentional system has to hold those pieces and analyze only a small subset at a time because of its limited capacity. Even if unlimited time is given decision may be less than perfect because the neural networks may not be able to put proper weightage on different subsets of variables it has examined serially over a long period of time.

In addition, conscious decisions are degraded if memory of past events is distorted. Because conscious processes analyze the pros and cons of each variable (past experiences, current situation, etc.) before making a decision, it is important that it has access to undistorted memory of past experiences. However, because our memory gets distorted in so many ways, it is virtually impossible to retrieve undistorted memory. In Chapter 2, Nonconscious memory, we discussed how memories of past events are distorted and how false memories are created. These distortions are described very elegantly by Daniel Schacter (Fig. 2.11) in his book *The Seven Sins of Memory*.[13]

Schacter describes seven common distortions: transience (forgetting over time); absent-mindedness (when memory is never encoded); blocking (temporary inaccessibility); misattribution (source of memory or event is forgotten or worse, misattributed); suggestibility (modification of memory based on suggestion); bias (memory modified by belief, mental state, or context); and persistence (certain memories refuse to go away and intrude on other memories). Because of these distortions conscious memory cannot always be relied upon for accuracy.

Because memories could be distorted in so many ways, it is not surprising that conscious decisions use "distorted" facts as important inputs. Forgetting is probably the most common cause of suboptimal conscious decisions. To make a good decision it is important to recall all relevant previous experiences, decisions made, and outcomes. Since it is an "open-ended recall" we are bound to forget many similar experiences while making decisions. It is also possible that we do not consciously recall any relevant experience because we know memories are forgotten over time. Some memories fade in a few seconds while others stay for years, either in their original or distorted forms. For example, after dialing a telephone number for the first time we tend to forget the number within seconds, unless it has emotional or personal value. Forgotten memories do not help us make right decision. Additionally, as discussed earlier attentional resources are needed to process conscious memory. If the attentional system is not fully activated memory of an event is not encoded and thus not available for making decisions. Even if the memory of an event is encoded and is not forgotten, there is a possibility that it cannot be retrieved while making a decision because of blocking. Since blocking of names and events is a common occurrence (particularly in elderly individuals), conscious decisions are less than perfect in many instances. Misattribution and false memories are even more problematic because they make us believe "facts" that are either imaginary or distorted.

Memory is distorted also because the brain has a tendency to "fill gaps." It uses "imagination" to nonconsciously fill forgotten pieces of the memory of an event. Even though the pieces that are filled never happened, they become an inseparable part of the true memory of an event. Therefore, imagined parts become "true memory" in conscious recollection. Sometimes memory is also distorted by suggestion. As discussed in Chapter 2, Nonconscious memory, this issue has been at the center of a number of lawsuits in which children were made to narrate stories of sexual abuse by repeated suggestive questioning. Distorted memory is also responsible for many cases of so-called "recovered" or "repressed" memories. Several women in the United States have claimed to have recovered memories of childhood sexual abuse. Some of their accusers were convicted based on those "memories." An American psychologist named Elizabeth Loftus who studied several cases of repressed "recovered" memory found that most of the stories were just "stories,"[14] even though accusers believed that they retrieved memories of real events. The conscious memory therefore is distorted more often than one might think. These distortions undermine the reliability of eye witness testimony in which our judicial system has so much faith.

Thus, contrary to the popular belief, memory is not an accurate replay of past events. It is recollection of events as perceived and interpreted by a person. Additionally, because some forms of memories last longer than others, they tend to skew decisions by retrieving only a specific kind of memory. For example, traumatic and emotional memories persist longer than other forms of memory.[15] These memories therefore, play a greater role in conscious decision making process. The quality of a conscious decision is affected by these distortions. Some of these distortions such as persistence of traumatic memories may have survival benefits.

Partly because of their resilience, emotional and traumatic memories are responsible for clinical symptoms of posttraumatic stress disorder (PTSD) in vulnerable individuals. Almost all decisions these individuals make are affected by those memories. As a result they tend to distrust people and are always fearful. If that is not enough these memories over a period of time get distorted and exaggerate the extent of trauma, making them feel even more helpless and traumatized.

Memories of past events are important not only for conscious but also for nonconscious decisions. But these decisions are based on memory retained nonconsciously. So, what about distortions in nonconscious memory? Is it resistant to distortions? Probably not, but it distorts to a much lesser extent. Implicit memory is known to last several years in the same form. In an interesting experiment psychologist David Mitchell asked volunteers to perform implicit and explicit memory tasks. They were asked to either name or recognize pictures presented briefly. These volunteers were unexpectedly called back after 17 years and shown those pictures and some novel pictures in fragmented form and asked to recognize them (as described in Chapter 2: Nonconscious memory and Fig. 2.10). Even though volunteers had no recollection of participating in the original experiment 17 years earlier, they made significantly fewer errors in recognizing the pictures shown earlier as compared to the novel pictures. It suggests that they had intact nonconscious memory of what they had learned 17 years ago.[16] Since this finding was replicated by other investigators, Larry Jacoby, who conducted some of the early experiments on nonconscious memory suggested that some forms of nonconscious memory alter people's perceptual processes permanently, they essentially remain undistorted forever.[17] This is in sharp contrast to conscious memory, which fades away or gets distorted in a relatively short period of time.

It is therefore generally believed that conscious memories are less stable, easily forgotten, and distorted more often. One of the problems with nonconscious memory discussed earlier is improper association as illustrated by the potato example. These associations can affect decisions and lead to long-lasting changes in behavior and cognition.

The examples of biases and preferences discussed at the beginning of this chapter suggest that nonconscious associations are not helpful. But that is not true because most of the time it associates relevant events. It may even have survival benefits. Thus, our liking for sweet and disliking for bitter food may have helped us survive. While wandering in forests our hunter–gatherer forefathers thrived on fruits and berries in addition to hunting. It may not be a coincidence that most fruits and berries with energy-packed carbohydrates have sweet taste and most alkaloids and other unsafe forest products are bitter.

Thus, the nonconscious memory does get distorted but the extent of distortion is significantly less than that of the conscious memory because distortions of nonconscious memories are passed on to conscious memory. Moreover, as discussed earlier our experiments suggest that conscious memory is dependent on nonconsciously retrieved information. If nonconscious information is distorted, conscious memory will also be distorted. Conscious memory therefore is distorted by distortions of nonconscious memory and also by its own distortions discussed earlier. That is why, conscious memory is less reliable than nonconscious memory.

It is clear that distortions in nonconscious memory affect conscious memory, but does it happen the other way around? There is no convincing evidence to suggest that distorted conscious memory affects nonconscious memory, but it is a possibility particularly if consciously learned false information is repeated over and over again. This phenomenon was seen in Nazi Germany. By repeating false information, Hitler's deputies made people believe them. Thus, if false facts are consciously repeated, those repetitions are interpreted as "facts" and encoded as such in the conscious as well as nonconscious mind. It is a form of source misattribution.[13,18] Once a false fact is incorporated in memory, it can distort nonconscious processes. This mechanism has not been experimentally verified and additional research is needed to support this assumption.

The influence of nonconscious processes on decision making has been studied by many investigators. In a comprehensive review[19] Richard Nisbett of the University of Michigan and Timothy Wilson of the University of Virginia described a number of experiments in which participation of nonconscious processes in conscious decisions was demonstrated. Many studies arguably suggest that all conscious decisions are made by the nonconscious mind which allows the conscious mind access only the operative part or the final decision.[20] The decision making process and information used to arrive at a decision never enter our conscious awareness. That is why it is not uncommon for us to make a decision and not know why we made it. At times, we "make up" a reason to convince ourselves that we made a reasoned conscious decision based on "facts."

Because decisions are made at the nonconscious level we are not aware of the processes that influence them. It was demonstrated experimentally by psychologists George Goethals and Richard Reckman of Williams College in Massachusetts. They asked volunteers to decide whether children should be provided buses to promote racial integrity. Their responses were recorded and in the second session volunteers who were in favor of buses were given arguments against it and those who did not favor were argued in its favor. A separate group was not given any argument. Later, when they were asked to recall their initial response, most volunteers who were given arguments against their position gave a wrong answer but those not argued against correctly stated their initial position.[21] Obviously, the argument given against the initial position not only affected their decision but also made them believe it was their original position. They were unaware of the process that helped change their original decision. That process remained nonconscious.

Further evidence of our lack of knowledge about the decision-making process comes from a classic experiment called the "two-cords puzzle" developed by a University of Michigan psychologist Norman Maier (1900–77) in 1931.[22] In this experiment Maier asked volunteers to tie together two cords hanging from the ceiling. The cords were placed at a distance and volunteers could not reach both cords while standing in one location. None of the volunteers could complete the task. Then Maier who was wandering in the room casually put one of the cords in motion. Within 45 seconds all volunteers were able to tie the cords by swinging them and catching when they were close to each other. When volunteers were asked how did they get the idea, no one mentioned Maier putting the cord in motion and made up their own explanation. A psychology professor who was one of the volunteers said, "having exhausted everything else the next thing was to swing it. I thought of swinging across a river." Obviously, the cue was Maier's action but that was never encoded consciously. Nonetheless, the conscious mind made up an explanation that was not true but convincing to the volunteer and others.

In another similar experiment Michael Storms and Richard Nisbett[23] gave placebo pills to insomniac volunteers and asked them to take them 15 minutes before going to bed. One group was told that the pill would enhance arousal by increasing heart rate, causing breathing irregularity, and increasing alertness, while the other group was told the opposite, that the pill would relax them, decrease heart rate, lower body temperature, and reduce alertness. On average, volunteers in the arousal group reported getting to sleep 28% quicker and those in the relaxation group took 42% longer to sleep. When asked about the reason, people said things like, "I am getting slung with my roommate" or "the exams are over." None of them mentioned that the pill might have affected their sleep. Even when explicitly asked about it, they denied that the pill had any effect. Some of them mentioned that they forgot about it. Obviously, the stated role of the pill had an effect on sleep pattern. The arousal group might have nonconsciously thought that their anxiety symptoms were due to the pills and the relaxation group may have nonconsciously overrated their anxiety. These data underscore lack of conscious knowledge of the real reasons for making decisions. In fact, they suggest that decisions are made nonconsciously and conscious processes have only a modifying influence. This modification makes decisions culturally and socially appropriate.

As mentioned earlier, emotions have a significant influence on our decisions. It may not be an exaggeration to say that many, if not all, of our decisions are influenced by emotions. This influence is highlighted in the famous "trolley problem," which was initially given to undergraduate students at the University of Wisconsin in 1905 by a philosophy professor Frank Chapman Sharp (Fig. 6.1).[24] The original "problem" was modified by a number of investigators and it has become one of the most popular experiments for demonstration of moral dilemma and the effect of emotion on decision-making process. In the modified version, the problem goes like this: Imagine a runaway trolley running on a track at a high speed. You realize that there are five people bound to the track and they are sure to die if the trolley continues to run on the track. You are standing in front of a lever that you can pull to divert the trolley to another track and save five people from certain death. However, you also notice that there is a person sitting on the other track where you would send the trolley by pulling the lever. In this situation if you do not do anything five people will die but if you change the track one person will die. On a rational level the decision should be easy: pull the lever and save five lives at the cost of one. But many people will have problem making that decision because they would not want to make a decision that would kill a person. In the original story that one person was the son of the individual who was to pull the lever. This problem highlights how emotion and moral values affect our decision. We do not make decisions based only on rationality. There are many factors that influence our decisions.

Because of their strong influence on the decision-making process, emotions are responsible for some of the most destructive and illogical decisions

FIGURE 6.1 Frank Chapman Sharp (1866–1943). Created the "trolley problem." *Image courtesy University of Wisconsin-Madison Archives.*

ever made by smart people. That is why we are told not to make emotional decisions. Emotions facilitate decisions that are mood congruent and undermine those that are mood incongruent. Thus, a sad person is more likely to decide to commit suicide than someone in a happy mood. Similarly, an angry individual is more likely to become aggressive than someone in a sad mood. The relationship between emotion and decision is reciprocal.[25] For example, a person fearful of flight would generate a fear response just before taking a flight and may decide not to take the flight, which would enhance this fear in the future. It is an example of immediate emotion. Emotions can also be anticipatory when they are triggered by anticipation of an event. For example, if a person buys a lottery ticket with the hope of winning a big prize she will get disappointed if the ticket does not win a prize. Emotion makes this person feel as if she lost the prize money. Anticipatory emotions are exploited in casinos to make people keep gambling even if they are losing. Anticipatory emotion triggered by small periodic wins make a gambler optimistic about future wins.

Immediate emotions are often more intense and drive people to make irrational decisions. For example, an individual who is fearful of flying may decide to use a riskier mode of transportation instead of flying. He may make an irrational decision because of intense somatic and autonomic symptoms associated with immediate emotion.

Despite having occasional negative outcome, autonomic reaction triggered by emotion has a survival benefit because it prepares us to deal with the situation that elicited the emotion. Thus, the fear response evoked by the sight of a gun-wielding maniac would prepare a person for fight or flight. To maximize chance of survival either by fighting or running away we need autonomic arousal that would increase heart rate, dilate pupils, cool off skin, and make us more energetic and better prepared to fight or flight. Emotions generated by that sight help make these autonomic changes.

An opposing view about the relationship between emotion and autonomic activation was proposed by an American-Hungarian neurologist Antonio Damasio. He suggested that emotions are initiated by the autonomic response elicited by an event/object.[26] However, this theory (the somatic marker theory) is counterintuitive and has no strong scientific support. According to the theory, somatic changes triggered by an event lead to emotional initiation. Thus, the sight of a gun-wielding maniac would elicit an autonomic response that would in turn initiate the emotion of fear. Sounds like putting the cart before the horse!

When emotionally guided decisions are different from those based on reason, our mind tries to find a middle ground. But usually emotions prevail. A good example of this is the Ultimatum Game, which was originally developed by a German economist Werner Guth (Fig. 6.2) and his colleagues in 1982.[27] This is a game that explains many processes in economics, psychology, philosophy, sociology, and neuroscience. Depending on the context it is modified and thus several versions of the game are available. A simple

FIGURE 6.2 Werner Guth (born 1944). Developed the Ultimatum Game. *Picture courtesy Prof Guth.*

version is played by two players, a proposer and a respondent. The proposer is given money to be divided between two players by making an offer to respondent and retain as much money as possible. If the respondent accepts the offer, he/she keeps the money offered by the proposer. However, if the respondent rejects the offer nobody gets money. In this game, the respondents often reject the offer that is considered unjust, even though rejection means he or she gets no money. Interestingly, respondents reject unjust offers only if the proposer is a live person. They accept "unjust" offers from a computer.[28] This game explains why we knowingly make bad and illogical decisions in life based on emotions. Similarly, almost all studies on voting behavior suggest that people vote on the basis of "emotional preference" rather than the policy position of a candidate. A candidate's look, body language, sincerity, and above all, ability to emotionally move voters, carry higher weight than his or her policies.[29]

How emotions influence our decisions is unclear because the brain network involved in the decision-making process has not yet been fully understood. It probably involves multiple networks depending on the nature of the decision and variables involved. Because of the involvement of multiple networks and variables, it takes time to make conscious decisions but nonconscious decisions are incredibly fast and at times almost instantaneous. It is not known how multiple variables are processed to make an instantaneous decision. It could however be possible if decisions are based on predefined templates or modules. The nonconscious mind may have a module for each component of the decision-making process. These modules may contain predefined sequence of neural activations leading to a predefined decision. By activating those modules the nonconscious mind could make a quick decision. The conscious decision-making process probably goes through several steps and may include exploration of possible options; weighing the pros and cons of each option; determining whether those options are consistent with cultural, moral, and social values; anticipating consequences of the decision; and evaluating whether the individual is capable of implementing it. There may be additional steps that conscious decision may have to go through. These steps make the conscious decision a slow process.

We have begun to understand how some of the steps of the decision making process are processed in the brain. As discussed in Chapter 5, Attention, we know "unwanted" options in the process are inhibited by executive or automatic inhibitory systems.[30,31] This is how the brain eliminates nonviable or imperfect options in the decision-making process. We also know that the anterior cingulate cortex (ACC) is involved in conflict resolution.[32–34] It helps the brain select the best option/response, among competing options. Neuroimaging studies have found activation in this area in tasks that require volunteers to select an option when multiple options are available. That is why the ACC is almost always activated in the Stroop task,[35,36] which requires volunteers to decide whether to read the word or name the color of ink as discussed earlier.

The cingulate is also involved in monitoring errors in the decision-making process. In a an experiment we observed activation in the cingulate in a task that required error monitoring and word generation.[37] In this experiment we had volunteers perform two decision-making tasks simultaneously. They had to generate a verb for a given noun and respond after waiting 1.5 seconds. Feedback about accuracy of the wait time was given after each response. The feedback allowed participants to "detect error." We were interested in understanding whether the cingulate cortex is involved in selecting an appropriate verb and also whether it monitors error to help improve response. We found activation in different areas of the cingulate during error detection and verb generation (Fig. 6.3). It was in the midcingulate region for error detection and in the anterior and posterior parts of the cingulate during verb generation. The anterior cingulate is also activated when a volunteer decides to turn down an "unjust" offer in the Ultimatum Game[38]. Interestingly, the level of cingulate activation correlated with the extent of

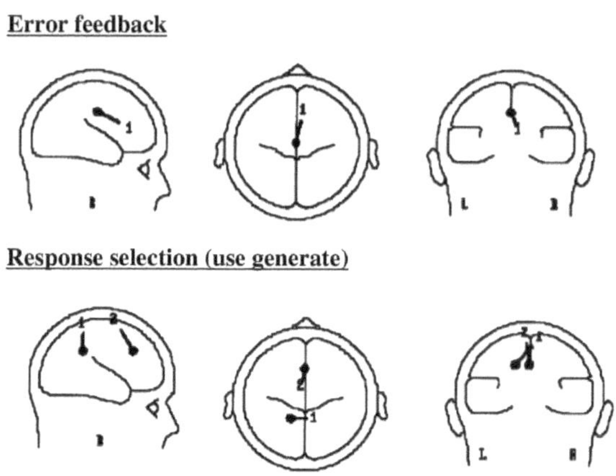

FIGURE 6.3 Different areas of the cingulate cortex were activated during verb generation and error detection tasks. *Reprinted from Badgaiyan RD, Posner MI. Mapping the cingulate cortex in response selection and monitoring. Neuroimage 1998;7(3):255–260.*

perceived injustice in the offer. It was least significant while playing with a computer.[28] The anterior cingulate activation therefore may be associated with selection of an appropriate response—a verb in our experiment and rejection of the offer in the Ultimatum Game.

Since emotion is an important component of the decision-making process, impairment of emotional processing significantly affects our decisions. This is why a lesion in the brain areas that control emotion lead to impaired decision. Thus, as discussed above individuals with a damaged ventral frontal cortex had difficulty making decisions in the Iowa gambling task[4] because of its involvement in emotional processing.[39] Lack of emotion and perhaps lack of nonconscious decision in people with a damaged prefrontal cortex lead to another interesting situation in the gambling task. These individuals despite having conscious knowledge that a deck is bad continue to pick cards from that deck, indicating lack of confidence in their own conscious knowledge/decision.[40] It could possibly be due to lack of emotion resulting from the frontal damage. If that were the case, emotions play an important role in building confidence in a decision.

A frequently cited example of disrupted decision-making process following prefrontal damage is one of the most frequently cited patients in medical literature. This patient was Phineas Gage. He was born in New Hampshire in July of 1823 and worked as a blasting foreman in a railway construction project. On September 13, 1848, while setting up explosives to blast a rock, he had an accident. The rock was blasted by making a hole and inserting in it a rod along with an explosive and sand. At around 4:30 pm that day, workers forgot to put sand in the hole and Gage got distracted by one of the workers who wanted to talk to him. As he turned his head the explosive fired and the rod shot up piercing Gage's head, entering his left jaw and coming out of his head (Fig. 6.4). The rod (110 cm long, 3.3 cm in diameter, and tapered at the end to 6 mm) pierced his left brain and landed about 80 ft away. Gage was thrown; he landed on his back and had convulsions. Within a few minutes he began talking and walking. He sat upright on an oxcart for a 1.2-km ride to the hotel where he lived. Within 30 minutes of the accident a local doctor named Edward William arrived to examine Gage. The doctor saw him sitting outside the hotel in a chair, describing the incidence. During the examination he vomited and expelled about half a cup of brain tissue. Despite this, Gage was fully conscious and joked about his condition. Realizing the severity of the injury, Dr. William transferred his care to a senior doctor J.M. Harlow about an hour later.[41,42] Harlow removed bone pieces and about 30 g of protruding brain tissue before putting a bandage over the wound. Gage was alert and conscious for the next three days, but then became semicomatose and spoke in monosyllables. His friends and family prepared his coffin and waited for him to die, but his condition started improving after the abscess in his head was drained. About two months after the accident he moved to his parents' house and started working on a farm.

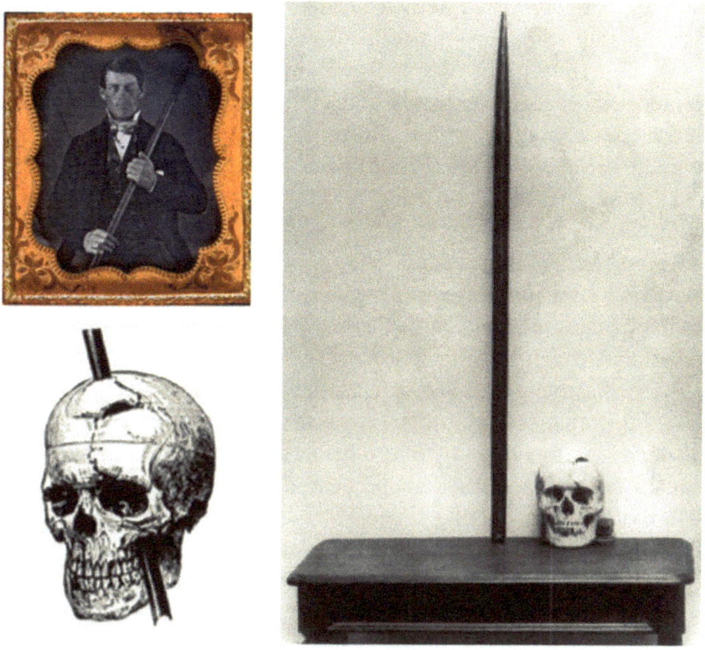

FIGURE 6.4 Phineas Gage (1823–1860). Skull and the rod that pierced Gage's brain is on display in the Countway Library, Harvard Medical School Boston. *Reprinted from Wikimedia Commons.*

At the time he was somewhat of a celebrity for surviving the accident. In August of 1852 Gage moved to Chile to work as a long-distance stagecoach driver. In 1959 his health began deteriorating. He therefore moved to San Francisco to live with his mother and sister and started working on a farm again. Gage used to have frequent epileptic seizures and on May 21, 1860 died of status epilepticus. He was buried in San Francisco. In 1866 when his physician Dr. Harlow knew of his death, he requested the family to exhume the body and donate his skull. The family agreed. His skull and the rod that pierced it are now available for public view in the Warren Museum in the Countway Medical Library of Harvard Medical School in Boston (Fig. 6.4).

Since Gage had achieved celebrity status, a lot has been written and said about his postaccident behavior. Despite abundance of information, it is unclear what were the consequences of the injury, because most of the information is clearly untrue. For example, it was said that he mistreated and abused his wife and children after the accident even though he was never married and had no children. His posttraumatic symptoms are thereforeunclear. Dr. Henry Jacob Bigelow who was the Professor of Surgery at Harvard Medical School studied Gage and described two major posttraumatic changes,[43] "... he had become gross, profane and vulgar" and that he was unable to make long-term decisions. His frequent change of job is cited as an example of the lack of decision-making ability. Antonio Damasio and others believe that his decision-making ability was compromised due to his damaged ventral prefrontal cortex.[26] It was suggested that the damage affected his emotion processing, which in turn impaired the ability to make long-term decisions. Even today Gage's disability is as controversial as it was during his lifetime.

Another brain area involved in emotional processing is the amygdala.[39] Damage to this area also impairs decision-making ability. A patient called SM, who has complete lesion of amygdala of both sides due to a rare genetic disorder called Urbach-Wiethe disease is discussed in Chapter 7, Emotion. Briefly, SM made many bad decisions in her life. Those decisions put her in life-threatening situations multiple times and almost killed her. Thus, inappropriate decisions made by people with damage to the prefrontal cortex or the amygdala, where emotions are processed, underscore the significance of emotion in the decision-making process.

It is clear that the nonconscious mind plays an important role in decision-making process. Impairment of nonconscious processing therefore significantly alters the quality of decision. Since most decisions are made nonconsciously, we are not even aware of the problem while making bad decisions. To make things worse, our conscious mind rationalizes those decisions. To make things worse, our conscious mind rationalizes those decisions. This rationalization over a period of time becomes part of our personality.

Bibliography

1. Dijksterhuis A, Bos MW, Nordgren LF, van Baaren RB. On making the right choice: the deliberation-without-attention effect. *Science* 2006;**311**(5763):1005–7.
2. Badgaiyan RD. Nonconscious perception, conscious awareness and attention. *Conscious Cogn* 2012;**21**(1):584–6.
3. Shanks DR, Newell BR, Lee EH, Balakrishnan D, Ekelund L, Cenac Z, et al. Priming intelligent behavior: an elusive phenomenon. *PLoS One* 2013;**8**(4):e56515.
4. Bechara A, Damasio AR, Damasio H, Anderson SW. Insensitivity to future consequences following damage to human prefrontal cortex. *Cognition* 1994;**50**(1–3):7–15.
5. Bechara A, Damasio H, Tranel D, Damasio AR. Deciding advantageously before knowing the advantageous strategy. *Science* 1997;**275**(5304):1293–5.
6. Bechara A, Tranel D, Damasio H. Characterization of the decision-making deficit of patients with ventromedial prefrontal cortex lesions. *Brain* 2000;**123**(Pt 11):2189–202.
7. Mamede S, Schmidt HG, Rikers RM, Custers EJ, Splinter TA, van Saase JL. Conscious thought beats deliberation without attention in diagnostic decision-making: at least when you are an expert. *Psychol Res* 2010;**74**(6):586–92.
8. Badgaiyan RD. Nonconscious processing and a novel target for schizophrenia research. *Open J Psych* 2012;**2**(4A). Available from: https://doi.org/10.4236/ojpsych.2012.224047.

9. Payne JW, Bettman JR. Walking with the scarecrow: the information-processing approach to decision research. In: Koehler DJ, Harvey N, editors. *Blackwell handbook of judgment and decision making*. Blackwell Publishing Ltd; 2004. p. 110–32.
10. Goldstein DG, Gigerenzer G. Models of ecological rationality: the recognition heuristic. *Psychol Rev* 2002;**109**(1):75–90.
11. Wilson TD, Schooler JW. Thinking too much: introspection can reduce the quality of preferences and decisions. *J Pers Soc Psychol* 1991;**60**(2):181–92.
12. Posner MI, Snyder CR, Davidson BJ. Attention and the detection of signals. *J Exp Psychol* 1980;**109**(2):160–74.
13. Schacter DL. *The seven sins of memory: how the mind forgets and remembers*. Houghton Mifflin; 2001.
14. Loftus E, Ketcham K. *The myth of repressed memory: false memories and allegations of sexual abuse*. St. Martin's Press; 1994.
15. Pillemer DB. *Momentous events, vivid memories*. Harvard University Press; 1998.
16. Mitchell DB. Nonconscious priming after 17 years: invulnerable implicit memory? *Psychol Sci* 2006;**17**(11):925–9.
17. Jacoby LL, Dallas M. On the relationship between autobiographical memory and perceptual learning. *J Exp Psychol General* 1981;**110**(3):306–40.
18. Schacter DL. The seven sins of memory. Insights from psychology and cognitive neuroscience. *Am Psychol* 1999;**54**(3):182–203.
19. Nisbett RE, Wilson TD. Telling more than we can know: verbal reports on mental processes. *Psychol Rev* 1977;**8/4**(3):231–59.
20. Miller G. Decision units in the perception of speech. *IRE Trans Inf Theory* 1962;**8**(2):81–3.
21. Goethals GR, Reckman RF. The perception of consistency in attitudes. *J Exp Soc Psychol* 1973;**9**:491–501.
22. Maier NRF. Reasoning in humans: Ii. The solution of a problem and its appearance in consciousness. *J Comp Physiol Psychol* 1931;**12**:181–94.
23. Storms MD, Nisbett RE. Insomnia and the attribution process. *J Pers Soc Psychol* 1970;**16**(2):319–28.
24. Sharp FC. Study of the influence of custom on the moral judgment. *Bull Univ Wisconsin* 1908;(236):138.
25. Lerner J, Keltner JR. Beyond valence: toward a model of emotion-specific influences on judgement and choice. *Cognition Emotion* 2000;**14**(4):473–93.
26. Damasio AR. *Descartes' error: emotion, reason, and the human brain*. G.P. Putnam; 1994.
27. Guth W, chmittberger R, Schwarze B. An experimental analysis of ultimatum bargaining. *J Econ Behav Organ* 1982;**3**(4):367–438.
28. Sanfey AG, Rilling JK, Aronson JA, Nystrom LE, Cohen JD. The neural basis of economic decision-making in the ultimatum game. *Science* 2003;**300**(5626):1755–8.
29. Westen D. The political brain: the role of emotion in deciding the fate of the nation. *PublicAffairs* 2007.
30. Casey BJ, Thomas KM, Welsh TF, Badgaiyan RD, Eccard CH, Jennings JR, et al. Dissociation of response conflict, attentional selection, and expectancy with functional magnetic resonance imaging. *Proc Natl Acad Sci USA* 2000;**97**(15):8728–33.
31. Badgaiyan RD, Sinha S, Sajjad M, Wack DS. Attenuated tonic and enhanced phasic release of dopamine in attention deficit hyperactivity disorder. *PLoS One* 2015;**10**(9):e0137326.

32. Badgaiyan RD. Executive control, willed actions, and nonconscious processing. *Hum Brain Mapp* 2000;**9**(1):38–41.
33. Badgaiyan RD. Nonconscious processing, anterior cingulate, and catatonia. *Behav Brain Sci* 2002;**25**:578–9.
34. Badgaiyan RD, Posner MI. Cingulate activation during use generation and error detection. *Neuroimage* 1997;**5**:S93.
35. Pardo JV, Pardo PJ, Janer KW, Raichle ME. The anterior cingulate cortex mediates processing selection in the stroop attentional conflict paradigm. *Proc Natl Acad Sci USA* 1990;**87** (1):256–9.
36. Lindsay DS, Jacoby LL. Stroop process dissociations: the relationship between facilitation and interference. *J Exp Psychol Human Perc Perf* 1994;**20**(2):219–34.
37. Badgaiyan RD, Posner MI. Mapping the cingulate cortex in response selection and monitoring. *Neuroimage* 1998;**7**(3):255–60.
38. Feng C, Luo YJ, Krueger F. Neural signatures of fairness-related normative decision making in the ultimatum game: a coordinate-based meta-analysis. *Hum Brain Mapp* 2015;**36** (2):591–602.
39. Badgaiyan RD, Fischman AJ, Alpert NM. Dopamine release during human emotional processing. *Neuroimage* 2009;**47**(4):2041–5.
40. Bechara A, Tranel D, Damasio H, Damasio AR. Failure to respond autonomically to anticipated future outcomes following damage to prefrontal cortex. *Cereb Cortex* 1996;**6** (2):215–25.
41. Harlow JM. Passage of an iron rod through the head. *Boston Med Surg J* 1848;**39**:389–93.
42. Harlow JM. Recovery from the passage of an iron bar through the head. *Publ Massachusetts Med Soc* 1868;**2**:327–47.
43. Bigelow HJ. Dr. Halow's case of recovery from the passage of an iron bar through the head. *Am J Med Sci* 1850;**20**(39):13–22.

Chapter 7

Emotion

As discussed in the previous chapter emotions help us make decision. They also encourage and motivate us to work to achieve our dreams and aspirations, making us creative. They are responsible, to a great extent for maintaining social order. Without emotion, we would be lacking empathy and killing each other at the slightest provocation. Imagine what would happen if parents had no emotional bonding with their children. They would neither provide protection nor help them grow. Under that scenario very few of us would have survived. Emotions therefore, are essential for our very survival. Probably that is why it is an involuntary function like other life-preserving activities. We have little or no voluntary control over its initiation or termination. However, with practice one can achieve limited control, the way we have limited voluntary control over breathing.

William James, who founded the modern psychology (Fig. 7.1), suggested that despite being controlled nonconsciously, emotion is consciously felt and experienced. In a paper titled "What is Emotion?"[1] James suggested that "Emotion is a perception of bodily changes. This perception forms a conscious feeling." He thought that feeling is an essential component of the emotional state and that it distinguishes emotion from other mental states. Sigmund Fraud (Fig. 1.6) also thought emotions originate nonconsciously and are expressed consciously,[2] but he also believed that under certain situations emotions can remain nonconscious. This idea was not accepted by others at the time. They strongly believed that emotion cannot remain nonconscious—it has to be experienced consciously.[3]

Experimental data support Freud's view and suggest that emotion can be nonconscious, not only in its origin but also in experience and expression. One of the experiments that suggest nonconscious nature of emotion was conducted in the laboratory of Ken Paller in Northwestern University. In this experiment volunteers were asked to rate their liking for emotionally neutral pictures (e.g., desk, chair). But before presenting pictures they were shown either happy or angry faces subliminally. Even though the subliminal faces were not consciously seen, volunteers' responses were influenced by those faces. As a result, they made a favorable response if a happy face preceded the picture and less favorable if it was preceded by an angry face.[4] Thus, even though faces were not consciously perceived, they elicited emotions,

Neuroscience of the Nonconscious Mind. DOI: https://doi.org/10.1016/B978-0-12-816115-9.00007-3

FIGURE 7.1 William James (1842–1910). Founder of the modern psychology. *Reproduced from Wikimedia Commons.*

FIGURE 7.2 Robert Zajonc (1923–2008). Described "mere exposure" effect. *Courtesy, University of Michigan.*

which altered volunteers' opinion of neutral pictures. Volunteers in this experiment were consciously unaware of their emotions that led to their favorable or unfavorable rating. They were also consciously unaware of the source of emotion. In another similar experiment conducted at the University of California San Diego, Piotr Winkelman and his colleagues subliminally presented happy or angry faces to volunteers. Immediately thereafter, they were given a drink to enjoy. After volunteers enjoyed the drinks, they were asked, how much they are willing to pay for it. Not so surprisingly, volunteers who were shown happy faces were willing to pay more than those who saw angry faces.[5] These experiments demonstrate that nonconsciously acquired information affects emotion and behavior without our conscious knowledge.

Another example that suggests nonconscious processing of emotion is so called mere-exposure effect[6] developed by a Polish-American psychologist Robert Zajonc (Fig. 7.2). Zajonc was born and raised in Poland during World War II. During the war his house was bombed killing his parents and seriously injuring him. After the bombing, he was sent to a German labor camp from where he escaped but was recaptured and sent to a political prison in France. He escaped again and enrolled in the University of Paris. In 1944 he moved to England and worked as a translator for American forces. At the end of war he immigrated to the United States and received PhD degree from the University of Michigan where he served as a professor until 1994. He then moved to Stanford University and died due to pancreatic cancer in December 2008.

Zajonc discovered the mere-exposure effect and demonstrated how people develop emotional connection with a stimulus just by its repeated exposure.[6] He extended this concept to nonconsciously presented stimuli by showing the emotional response to stimuli that were not recognized consciously.[7] He also demonstrated that people develop a liking for even nonsense words like *zebbilons* and *warbus* after repeated exposure. To prove that a positive emotion develops nonconsciously just by being familiar with a stimulus, Zajonc asked volunteers to look at 25 pictures presented subliminally for 5 ms. One group was shown 25 different pictures while the other was shown 5 pictures that were repeated 5 times. After the session, it was found that volunteers of the second group were in positive affective state even though they did not perceive the stimuli consciously. Interestingly, volunteers did notice a change in their emotional state but had no idea why it changed.[8] The experiment demonstrated that nonconsciously perceived stimuli elicit emotional response that we are not consciously aware of.[9]

Even stronger evidence in support of nonconscious emotion comes from the study of infants born with a congenital condition called anencephaly. These babies are born without a cerebral cortex and as a result lack awareness. Despite this, anencephalic children show positive facial expressions when given sugar and the negative emotion of disgust when a bitter substance is put on their tongue.[10] Similarly, animals like fish and reptiles are known to show emotional reactions even though they do not have evolved consciousness. These observations prompted John Kihlstrom, a psychologist at the University of California Berkeley, to propose two forms of emotion, explicit and implicit. We are consciously aware of only explicit emotion. Implicit emotion remains outside our conscious awareness.[11]

Experiments that endorse the existence of nonconscious emotion include a study conducted by Ulf Dimberg and colleagues[12] of Uppsala University. They recorded activity of facial muscles while presenting subliminal happy or angry faces to volunteers. The study suggested that subliminally presented faces activate the same muscles that are active when looking at visible happy or angry faces. Additionally, subliminal stimuli elicit the galvanic skin response (GSR)[13] indicating anxiety and activate brain areas that process emotions.[14,15] These areas include the amygdala, medial temporal lobe (MTL), and ventral prefrontal cortex.[16]

The experiments described above suggest that stimuli that cannot be consciously perceived do initiate neural changes associated with emotional

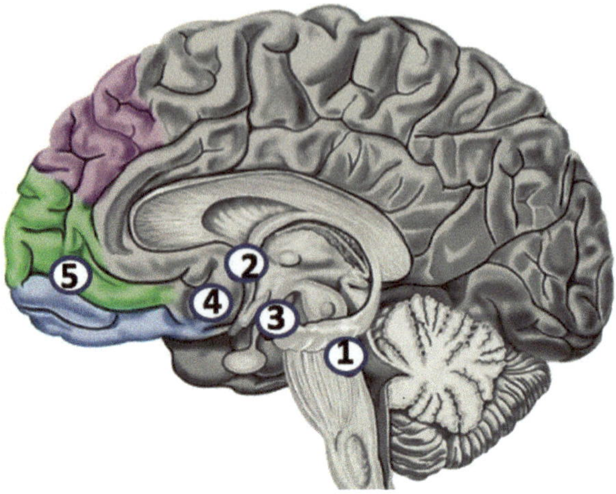

FIGURE 7.3 Location of the brain areas involved in emotional processing: (1) parabrachial nucleus; (2) nucleus accumbens; (3) amygdala; (4) medial temporal lobe; and (5) ventral prefrontal cortex.

arousal. But it is not clear whether volunteers in these experiments were subjectively aware of the elicited emotions. If they were, it would support William James' idea that emotions must be experienced consciously.[1]

Incidentally, many of the brain areas involved in processing of emotion are essentially areas that mediate nonconscious functions. Thus, emotions are initiated in the parabrachial nucleus located in the brain stem that controls functions like breathing and blood circulation (Fig. 7.3). The parabrachial nucleus receives input from almost all sensory modalities. Based on this input it creates the basic framework of emotion to be generated. At this level emotion is not well defined. Only a template for either positive or negative emotion is created. This template is modified and refined in the higher brain areas. The parabrachial nucleus decides the type and intensity of emotion. That is why injection of certain drugs like benzodiazepines into this area enhances positive emotions.[17] Based on sensory signals it receives from different parts of the body, the parabrachial nucleus generates emotions.[18] Signals from the parabrachial nucleus are further processed in the nucleus accumbens (Figs. 7.3 and 7.4), which is a part of the basal ganglia and strategically located to allow communication with both lower and higher brain areas. Because of its connections with the higher brain areas, its output eventually makes us consciously aware of the emotion. Until signals are relayed from the nucleus accumbens to the cerebral cortex, they remain nonconscious.

In addition to emotion, the nucleus accumbens is also involved in reward processing, which controls addictive behavior.[19] That is why both reward and addiction modify emotions. The reward system evaluates stimuli and assigns emotional values based on their rewarding potential. Addiction also influences emotion by activating the reward system. The nucleus accumbens

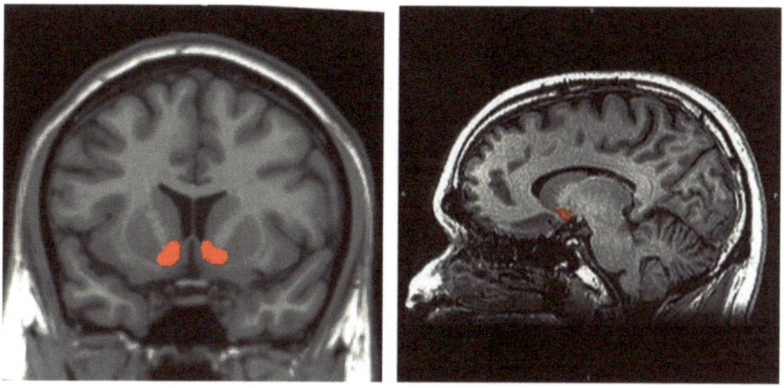

FIGURE 7.4 Location of the nucleus accumbens (red) in coronal (left) and sagittal (right) sections of the brain. It binds emotion, addiction, and the reward system as discussed in the text.

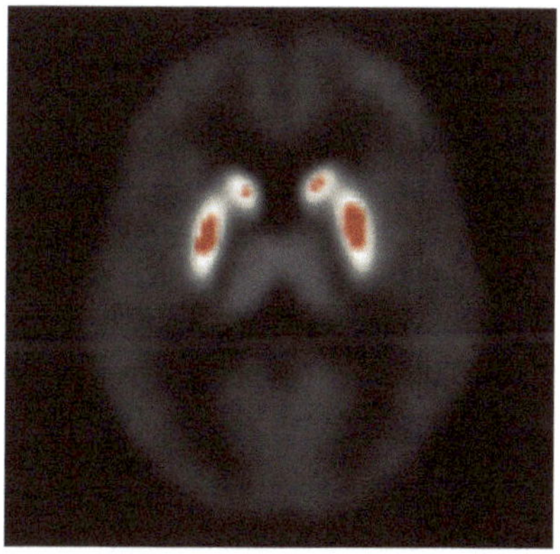

FIGURE 7.5 Dopamine release in the caudate and putamen during emotional processing (*red*) detected and mapped using the neurotransmitter imaging technique SDMIT. *Adapted from Badgaiyan RD. Dopamine is released in the striatum during human emotional processing.* Neuroreport *2010;**21**(18):1172–6.*

therefore binds together emotion, reward, and addiction. Based on the reward potential of the emotional signals relayed from the parabrachial nucleus the nucleus accumbens modifies and calibrates intensity of the emotion. That is why stronger emotions cause greater activation of the nucleus accumbens. In the human brain emotion is processed both in the ventral and dorsal striatum (the striatum is a part of the basal ganglia). The nucleus accumbens is located in the ventral striatum, and the dorsal striatal structures involved in the processing are the caudate and putamen. Because of their involvement we found increased release of dopamine in the caudate and putamen when emotions were elicited in healthy volunteers (Fig. 7.5).[20] To measure dopamine release in the live human brain during emotional processing, we used a novel neurotransmitter imaging technique called the single-scan dynamic molecular imaging technique (SDMIT).[21–24] As mentioned earlier, the technique allows detection, mapping, and measurement of the amount of dopamine released in the brain acutely during performance of a cognitive, emotional, or behavioral task.

From the striatum signals are relayed to the higher brain areas for further processing of emotion. In an experiment we demonstrated that this relay is also mediated by dopamine. Emotional signals activate the dopamine system in the amygdala, MTL, and ventral prefrontal cortex. In a SDMIT study we observed increased dopamine release in these areas during emotional processing (Fig. 7.6).[16] Interestingly, dopamine release was observed only in the left hemisphere during emotional processing. Thus, the amygdala, MTL, and frontal cortex of only the left side were activated in this experiment. It is not clear why it was released unilaterally. It may be due to low intensity of emotion elicited by brief presentation of emotional words in this experiment. Since words do not elicit strong emotion there was unilateral activation. It is possible that stronger emotions activate these areas bilaterally.

In these neurotransmitter imaging experiments using SDMIT,[20] we compared the amount of dopamine release while healthy young volunteers were looking at either emotionally neutral (e.g., park. pencil) or emotional (e.g., blood, rape) words on a computer monitor. Before they were shown words, a radiolabeled dopamine receptor ligand was injected intravenously. Because the ligand and dopamine bind to the same receptors in the brain, endogenously released dopamine displaces the ligand from the receptor site reducing its concentration. Therefore, by identifying brain areas where ligand concentration was reduced, we were able to map dopamine release during task performance.[23,24] The ligand concentration was measured using a positron emission tomography (PET) camera. By comparing concentration during presentation of neutral and emotional words, the amount and location of dopamine released during emotional processing was estimated. In these experiments we used two ligands to detect dopamine. It was detected in the striatum using a low-affinity ligand ^{11}C-raclopride and a high-affinity ligand ^{18}F-fallypride was used for detection of dopamine outside the striatum.[16] We had to use different ligands to detect dopamine inside and outside the striatum because of the differences in dopamine receptor density. It is high inside and low outside the striatum. Since raclopride cannot be detected in low receptor density areas and fallypride has limited displaceability in high receptor density areas of the brain, we had to conduct separate experiments using different ligands to detect dopamine inside and outside the striatum.

Out of the three brain areas where we observed dopamine release during emotional processing (Fig. 7.6), the amygdala is the most important. The primary role of this pear-shaped subcortical structure (Fig. 7.7) is to define emotion and retain memory of emotional events. The amygdalae of the two hemispheres appear to have slightly different function. Stimulation of the right amygdala generates negative emotions of fear, and sadness while left amygdala generates positive emotion of happiness.[25] Because of its connection to the MTL (specifically the hippocampus and parahippocampus), the

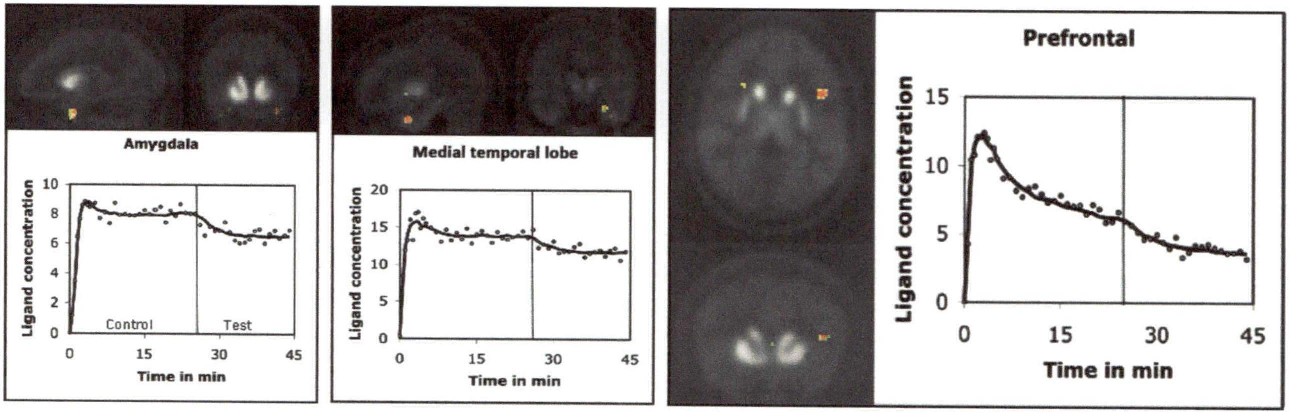

FIGURE 7.6 Dopamine release in the amygdala, medial temporal lobe, and ventral frontal cortex during emotional processing. Time activity curves show significant reduction in the ligand concentration in these areas when volunteers were shown emotional words (test condition) instead of neutral words (control condition). Reduced ligand concentration indicates release of endogenous dopamine in these areas. *Adapted from Badgaiyan RD, Fischman AJ, Alpert NM. Dopamine release during human emotional processing.* Neuroimage *2009;**47**(4):2041–5.*

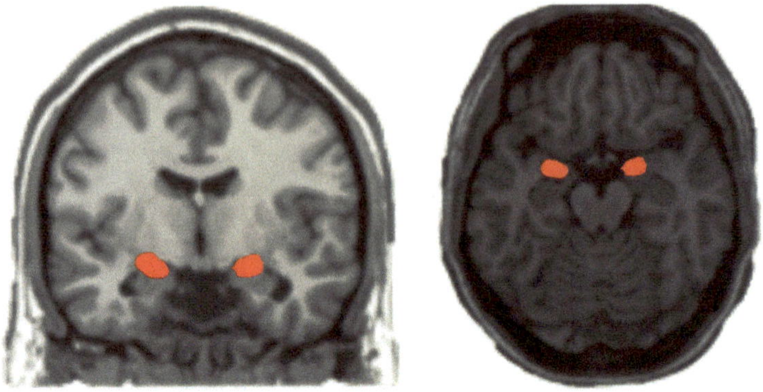

FIGURE 7.7 Location of human amygdalae (*red*) in coronal and sagittal sections of the brain.

amygdala provides a context to the emotion originally formulated in the parabrachial nucleus and modified in the striatum. Our data suggest that there is a dense dopaminergic connection between the amygdala and the MTL (primarily the parahippocampus gyrus).[16] Since MTL is an important area for memory processing, this connection allows the amygdala to fine tune emotions based on the past experience.

Even though the amygdala plays an important role in emotional processing, its precise function is unclear. Most of the data on its role in human emotional processing has been obtained by studying patients with a rare genetic disease called Urbach–Wiethe disease. In this condition, the amygdala is damaged bilaterally. One of those patients known as SM (also known as SM-046) was studied extensively by Antonio Damasio[26] in Iowa. Because of a damaged amygdala SM had no emotion. In particular, she lacked fear and anxiety. Horror movies and horror houses did not evoke any fear in her. Interestingly, despite claiming that she was afraid of snakes, when taken to a pet store she had no problem handling and fondling snakes. She even touched the tongues of snakes without any hesitation. Because of her lack of fear she often put herself in life-threatening situations and was mugged more than once. Even in those situations she had no fear or anxiety and did not avoid visiting the spots where she was assaulted. Additionally, she was almost killed in a domestic violence incident and received several death threats.

SM does not avoid potentially life threatening situations because she lacks fear but is very social in her interaction and has no anxiety meeting strangers or having physical contact with them. She frequently invades personal space and acknowledges that other people need more space than she does. SM also has difficulty judging the emotional expressions of other people. As a result, she does not know who to trust. Surprisingly, she has retained at least one fear response—the fear of being suffocated. She experiences fear while breathing air with high carbon dioxide concentration. Retention of this fear response indicates that emotions that are necessary for survival are not processed at the amygdala. These emotions probably remain unchanged after they are generated at the parabrachial nucleus of the brain stem (Fig. 7.3).

Memories of emotional events are processed both in the MTL and amygdala. Probably because of this dual processing we have better recollection of emotional memories. As expected, this advantage is lost if the amygdala is damaged. SM therefore did not show superiority in either explicit or implicit memory of emotional events. Her memory for emotional events is poor but recollection of nonemotional events is unaffected.[27] It appears that the amygdala helps encode emotional memories. If it is damaged later in life, previously encoded emotional memories remain intact.[28] The amygdala therefore is needed for encoding emotional memories but not for their retention or retrieval.

The amygdala and MTL work closely together to process emotion and to enhance encoding of emotional memories using the dopaminergic connection between them.[16] This connection establishes a channel for two-way exchange of information during emotional processing. Since the MTL is a part of the memory system,[29,30] it provides context for emotion using memories of past events. Based on this context emotions are further refined and calibrated in the amygdala. Motivational values assigned by the nucleus accumbens to an emotion are also calibrated at the amygdala,[31] which refines all emotions except those that are essential for survival.

Observations of impaired emotional processing in people with a damaged amygdala are consistent with findings of a classical experiment conducted in 1939 on monkeys by German-born American physiological psychologist Heinrich Kluver and American neurosurgeon Paul Bucy.[32] A chance finding of the two turned out to be an important discovery about functions of the temporal lobe. While studying the psychedelic drug mescaline, they removed the temporal lobe along with the amygdala of a monkey but did not find expected results. However, they noticed that the animal developed a triad of symptoms: placidity, omniphagia (eating everything edible as well as nonedible objects), and hypersexuality. The placidity was probably because of the amygdala lesion. Even though these symptoms were first described[33] in 1888 by a British physiologist Sir Edward Albert Schafer (1850–1935) and the American physician Sanger Brown (1852–1928), Kluver and Bucy's experiment was more popular, which is why the syndrome bears their names (Kluver-Bucy syndrome). Incidentally, Sir Schafer is better known for coining the word "endocrinology" and "insulin" and Brown was the first to demonstrate that the "center of vision" in monkeys is located in the occipital lobe.

In addition to the amygdala and MTL, the orbitofrontal cortex (OFC) located just above the eye sockets in the ventral frontal lobe (Fig. 7.8) is also involved in emotional processing.[16] Along with the amygdala this area refines emotions. Damage to the OFC therefore, causes some of the same impairments seen following damage to the amygdala.[26] Because of this reason,

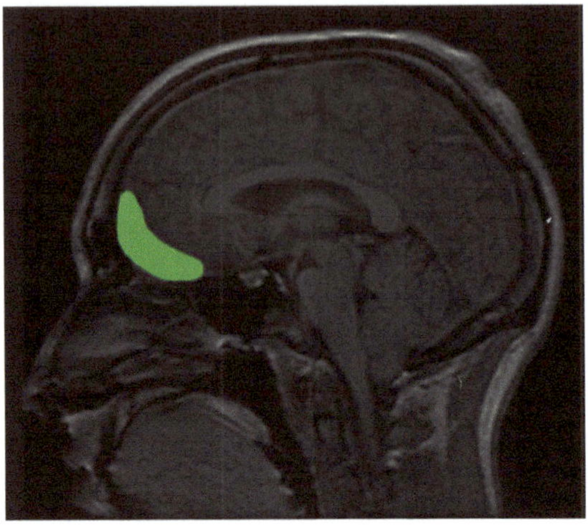

FIGURE 7.8 Location of the orbitofrontal cortex (*green*) involved in emotional processing. *Reproduced from Wikipedia Commons.*

Edmund Rolls of Oxford University proposed that in primates the OFC takes over functions of the amygdala and is the central area for emotional processing.[34] Even though the role of amygdala in human emotional processing cannot be discounted, the OFC has a significant modulatory influence. Emotional regulation in this area is more precise than that at the amygdala because it has access to memories of the past events and to all sensory modalities. Based on those inputs it calibrates affect and intensity of emotion.

As discussed earlier, the famous patient Phineas Gage had a damaged ventral prefrontal cortex, that included the OFC (Fig. 6.4). Because of this damage, it is thought that he was unable to make long-term decisions.[26] It is not clear whether Gage had impaired long-term decision-making ability but experiments conducted using the Iowa gambling task do suggest that individuals with a damaged OFC have impairment of both emotional processing and decision-making ability.[35] In this task individuals with a damaged ventral frontal cortex (including OFC) had difficulty making nonconscious decision about the nature of deck of cards. Additionally, unlike healthy volunteers they did not show anticipatory galvanic skin response (GSR) when cards were picked from bad decks, indicating lack of anxiety response. The integrity of this part of the brain therefore appears to be essential for making nonconscious decision and also for generating anxiety response. As discussed in Chapter 6, Decision making, emotion is an essential part of the decision-making process. The impaired decision-making ability of people with a damaged OFC could therefore be due to impaired emotional processing.

How emotion influences decision-making ability is unclear. As discussed in Chapter 6, Decision making, the nonconscious mind probably uses modules for making decisions. It probably builds a module for each significant event and each module includes a response to a specific situation. When the same situation occurs again, the brain does not need to go through all processing steps. It instead activates the relevant module that generates immediate response. Thus, if we encounter an unpleasant event, a response to avoid similar situation in the future is formulated in the module and on its activation a response is executed nonconsciously.[36] It appears that the OFC provides emotional input to these modules. It explains the dopaminergic connection between the OFC and the basal ganglia, which create the modules. In a series of neurotransmitter imaging experiments, we found dopamine release both in the ventral prefrontal cortex that includes the OFC and the basal ganglia during emotional processing.[16,20]

Another relevant observation came from the study of individuals with a damaged frontal cortex. In the Iowa gambling task these patients did not pick cards from good decks even though they knew which decks were good.[37] This behavior is surprising but not entirely unexpected. Occasionally we do act against our own decisions, particularly when we lack confidence. It therefore appears that individuals with a damaged prefrontal cortex lack confidence in their decisions. Since these patients are also unable to make nonconscious decision and lack emotional input from the OFC, both of these factors may play a role in diminishing their level of confidence.

Additionally, the OFC, along with the adjoining anterior cingulate cortex (ACC; Fig. 7.9), is responsible for assigning affective value to emotional content. As discussed earlier, the cingulate cortex is involved in many tasks and each task is performed by different cluster of neurons located in different parts of the cingulate.[38] The dorsal ACC is involved in cognitive processing and is connected to the prefrontal and parietal cortex. The ventral part receives input from the amygdala, nucleus accumbens, and hippocampus. It mediates emotion. Additionally, different clusters of neurons within the ventral ACC process different types of emotion. Neurons that process positive emotions (e.g., love, happiness) are located more ventrally than those that process negative emotions (e.g., fear, anxiety).

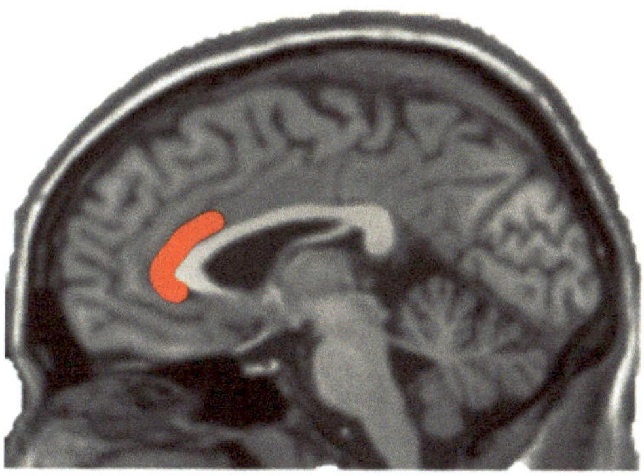

FIGURE 7.9 Location of the anterior cingulate cortex (ACC) in the human brain.

It is suggested that emotion is regulated in the OFC and ACC in two phases: evaluation and regulation.[39] Neurons located on the dorsal aspects of these structures evaluate emotional signals and those on the ventral aspect regulate them by assigning appropriate affective value. It is therefore not surprising that people with mood disorders have reduced volume of ACC[40,41] and have been successfully treated by surgical removal of the part of the ACC called the subgenicular part located in its anterior and ventral aspect.

Studies suggest that positive and negative emotions are preferentially processed in the prefrontal cortex of different hemispheres. The left hemisphere processes the positive and the right negative emotions. This is why people with clinical depression have reduced metabolic activity in the left prefrontal cortex. Reduced metabolic activity is indicated by a reduction in the amount of radiolabeled ^{18}F-dextroglucose utilization.[42] The reduced left prefrontal activity in clinical depression suggests that it could be resolved if activity of this area is enhanced. It indeed happens in some individuals when the left frontal cortex is activated using recurrent transcranial magnetic stimulation technique.[43]

Thus, emotion is regulated by multiple brain areas and most of the processing remains outside conscious awareness. This is why we have limited control over emotions and on occasion we are not even aware of the reason for a specific emotion. Because of its nonconscious nature, we occasionally experience sadness for no apparent reason or happiness without a clear cause. Because of its nonconscious nature it difficult, if not impossible to terminate emotions at will. Emotions can be modulated better using techniques that alter the nonconscious mind rather than the conscious mind.[44] Some of these techniques, including psychotherapy, are used therapeutically to treat clinical depression.

Even though emotions remain largely outside voluntary control, they do play an important role in motivating (or discouraging) us to perform voluntary actions. They also keep us safe by eliciting fear and anxiety responses, which is why people with impaired emotional processing often have chaotic and dangerous lives.[26]

Emotions have played a pivotal role in development of our civilization by creating passion leading to invention and innovation. They generate feelings of belonging and help create families and communities. If we had no emotion, we would probably not have survived that long as a species. At the very least we would not have been as creative as we have been throughout our civilization.

Bibliography

1. James W. What is an emotion. *Mind* 1884;**9**:188–205.
2. Freud S. The unconscious. In: Strachey J, editor. *The standard edition of the complete psychological works of Sigmund Freud*, 14. Hogarth Press and Institute of Psychoanalysis; 1915. p. 59–204.
3. Clore GL. Why emotions are never unconscious. In: Ekman P, Davidson RJ, editors. *The nature of emotion: fundamental questions*. Oxford University Press; 1994. p. 285–90.

4. Li W, Zinbarg RE, Boehm SG, Paller KA. Neural and behavioral evidence for affective priming from unconsciously perceived emotional facial expressions and the influence of trait anxiety. *J Cogn Neurosci* 2008;**20**(1):95–107.
5. Winkielman P, Berridge KC, Wilbarger JL. Unconscious affective reactions to masked happy versus angry faces influence consumption behavior and judgments of value. *Pers Soc Psychol Bull* 2005;**31**(1):121–35.
6. Zajonc RB, Reimer DJ, Hausser D. Imprinting and the development of object preference in chicks by mere repeated exposure. *J Comp Physiol Psychol* 1973;**83**(3):434–40.
7. Moreland RL, Zajonc RB. Is stimulus recognition a necessary condition for the occurrence of exposure effects? *J Pers Soc Psychol* 1977;**35**(4):191–9.
8. Monahan JL, Murphy ST, Zajonc RB. Subliminal mere exposure: specific, general, and diffuse effects. *Psychol Sci* 2000;**11**(6):462–6.
9. Berridge KC, Winkelman P. What is an unconscious emotion? (the case for unconscious "liking"). *Cognition Emotion* 2003;**17**:181–211.
10. Steiner JE. The gustofacial response: observation on normal and anencephalic newborn infants. *Symp Oral Sensory Perc* 1973;(4):254–78.
11. Kihlstrom JF. The psychological unconscious. In: Pervin LA, John OP, editors. *Handbook of personality: theory and research*. New York: Guilford Press; 1999.
12. Dimberg U, Thunberg M, Elmehed K. Unconscious facial reactions to emotional facial expressions. *Psychol Sci* 2000;**11**(1):86–9.
13. Öhman A, Flykt A, Lundqvist D. Unconscious emotion: evolutionary perspectives, psychophysiological data and neuropsychological mechanisms. In: Lane RD, Nadel L, Ahern G, editors. *Cognitive neuroscience of emotion*. Oxford University Press; 2000. p. 296–327.
14. Dolan RJ, Fletcher P, Morris J, Kapur N, Deakin JF, Frith CD. Neural activation during covert processing of positive emotional facial expressions. *Neuroimage* 1996;**4**(3 Pt 1):194–200.
15. Morris JS, Ohman A, Dolan RJ. Conscious and unconscious emotional learning in the human amygdala. *Nature* 1998;**393**(6684):467–70.
16. Badgaiyan RD, Fischman AJ, Alpert NM. Dopamine release during human emotional processing. *Neuroimage* 2009;**47**(4):2041–5.
17. Soderpalm AH, Berridge KC. The hedonic impact and intake of food are increased by midazolam microinjection in the parabrachial nucleus. *Brain Res* 2000;**877**(2):288–97.
18. Damasio AR, Grabowski TJ, Bechara A, Damasio H, Ponto LL, Parvizi J, et al. Subcortical and cortical brain activity during the feeling of self-generated emotions. *Nat Neurosci* 2000;**3**(10):1049–56.
19. Koob GF, Le Moal M. Drug addiction, dysregulation of reward, and allostasis. *Neuropsychopharmacology* 2001;**24**(2):97–129.
20. Badgaiyan RD. Dopamine is released in the striatum during human emotional processing. *Neuroreport* 2010;**21**(18):1172–6.
21. Badgaiyan RD, Fischman AJ, Alpert NM. Striatal dopamine release during unrewarded motor task in human volunteers. *Neuroreport* 2003;**14**(11):1421–4.
22. Badgaiyan RD. Neurotransmitter imaging: basic concepts and future perspectives. *Curr Med Imaging Rev* 2011;**7**:98–103.
23. Badgaiyan RD. Detection of dopamine neurotransmission in "real time". *Front Neurosci* 2013;**7**:125.
24. Badgaiyan RD. Imaging dopamine neurotransmission in live human brain. *Prog Brain Res* 2014;**211**:165–82.

25. Lanteaume L, Khalfa S, Regis J, Marquis P, Chauvel P, Bartolomei F. Emotion induction after direct intracerebral stimulations of human amygdala. *Cereb Cortex* 2007;**17**(6):1307–13.
26. Damasio AR. *Descartes' error: emotion, reason, and the human brain.* G.P. Putnam; 1994.
27. Adolphs R, Tranel D, Buchanan TW. Amygdala damage impairs emotional memory for gist but not details of complex stimuli. *Nat Neurosci* 2005;**8**(4):512–18.
28. Brierley B, Medford N, Shaw P, David AS. Emotional memory and perception in temporal lobectomy patients with amygdala damage. *J Neurol Neurosurg Psychiatry* 2004;**75**(4):593–9.
29. Badgaiyan RD. Neuroanatomical organization of perceptual memory: an fMRI study of picture priming. *Hum Brain Mapp* 2000;**10**(4):197–203.
30. Badgaiyan RD, Posner MI. Time course of cortical activations in implicit and explicit recall. *J Neurosci* 1997;**17**(12):4904–13.
31. Cardinal R, Parkinson J, Hall J, Everitt B. Emotion and motivation: the role of the amygdala, ventral striatum, and prefrontal cortex. *Neurosci Biobehav Rev* 2002;**26**(3):321–52.
32. Kluver H, Bucy PC. Preliminary analysis of functions of the temporal lobes in monkeys. 1939. *J Neuropsych Clin Neurosci* 1997;**9**(4):606–20.
33. Brown S, Schafer EA. An investigation into the functions of the occipital and temporal lobes of the monkey's brain. *Philosoph Trans Royal Soc London B* 1888;**179**:303–27.
34. Rolls ET. Functions of the orbitofrontal and pregenual cingulate cortex in taste, olfaction, appetite and emotion. *Acta Physiol Hung* 2008;**95**(2):131–64.
35. Bechara A, Damasio AR, Damasio H, Anderson SW. Insensitivity to future consequences following damage to human prefrontal cortex. *Cognition* 1994;**50**(1–3):7–15.
36. White NM. Mnemonic functions of the basal ganglia. *Curr Opin Neurobiol* 1997;**7**(2):164–9.
37. Bechara A, Tranel D, Damasio H, Damasio AR. Failure to respond autonomically to anticipated future outcomes following damage to prefrontal cortex. *Cereb Cortex* 1996;**6**(2):215–25.
38. Badgaiyan RD, Posner MI. Cingulate activation during use generation and error detection. *Neuroimage* 1997;**5**:S93.
39. Atkins AS, Reuter-Lorenz PA. Neural mechanisms of semantic interference and false recognition in short-term memory. *Neuroimage* 2011;**56**(3):1726–34.
40. Drevets WC, Ongur D, Price JL. Neuroimaging abnormalities in the subgenual prefrontal cortex: implications for the pathophysiology of familial mood disorders. *Mol Psych* 1998;**3**(3):220–6 190–221.
41. Mayberg HS, Brannan SK, Mahurin RK, Jerabek PA, Brickman JS, Tekell JL, et al. Cingulate function in depression: a potential predictor of treatment response. *Neuroreport* 1997;**8**(4):1057–61.
42. Baxter Jr. LR, Schwartz JM, Phelps ME, Mazziotta JC, Guze BH, Selin CE, et al. Reduction of prefrontal cortex glucose metabolism common to three types of depression. *Arch Gen Psych* 1989;**46**(3):243–50.
43. Brunelin J, Poulet E, Boeuve C, Zeroug-vial H, d'Amato T, Saoud M. Efficacy of repetitive transcranial magnetic stimulation (rTMS) in major depression: a review. *Encephale* 2007;**33**(2):126–34.
44. Williams LM, Mathersul D, Palmer DM, Gur RC, Gur RE, Gordon E. Explicit identification and implicit recognition of facial emotions: I. Age effects in males and females across 10 decades. *J Clin Exp Neuropsychol* 2009;**31**(3):257–77.

Chapter 8

Creativity

Creativity is perhaps the single most important function that has enabled the human race to dominate the world. Like other nonconscious functions, we have little or no voluntary control over creativity. Creativity comes from "deeper processing" of information. Geniuses therefore expand on the range of options available to accomplish a task and find options that most people do not think of. It allows creative people to make novel associations and discoveries. The ability to make unusual and novel associations is an important quality of creative thinkers.

Theoretically, our brain will make unusual associations if the neural networks are altered. An altered network allows thought processes to go in the direction it normally does not go. We see this kind of behavior in people with psychiatric conditions. Because of unusual neural networks these individuals process information differently and therefore have altered cognition and behavior. Since these alterations are usually maladaptive, they have limited ability to meet professional and social obligations. These alterations, however, also provide them the ability to make unusual associations and think "outside of the box." That is why many people with mental illnesses are extremely creative.

The link between mental illness and creativity is known for centuries. It was documented by Greek philosopher Aristotle (Fig. 8.1) in one of his classic writings *Problemata*. He mentioned the frequent occurrence of "melancholia" in prominent people of the time. Later, Cesare Lombroso (1835–1909), an Italian physician considered the father of criminology, studied the biographies of prominent people and concluded that psychiatric illness, creativity, and addiction tend to co-occur. In recent years, Nancy Andreasen, a National Medal of Science awardee psychiatrist at the University of Iowa, followed 30 writers for 15 years.[1] She found that 80% of them had some form of mood disorder. Most of them (44%) were diagnosed with bipolar disorder, which is characterized by mood swings between low (depression) and high (mania or hypomania). Andreasen's findings are consistent with observations of others studies. While there is a strong correlation between creativity and mood disorders, support for a similar correlation with

Neuroscience of the Nonconscious Mind. DOI: https://doi.org/10.1016/B978-0-12-816115-9.00008-5

FIGURE 8.1 Aristotle (384–322 BCE) made the connection between creativity and mental illness. *Reprinted from Wikimedia Commons.*

FIGURE 8.2 John Nash (1928–2015). Nobel prize-winning mathematician who carried a diagnosis of paranoid schizophrenia. *Reprinted from Wikimedia Commons.*

psychotic conditions like schizophrenia is not as strong, even though ancient thinkers believed that psychotic people are creative. It does not mean that people with psychotic disorders are not creative. Many of them are. Throughout history there are examples of extraordinary talent in people with psychotic symptoms. Perhaps the best-known example of our time is the Princeton mathematician and Nobel laureate John F. Nash (Fig. 8.2). He is considered the father of game theory, which defines how people, animals, and computers make rational decisions. In 1994 he was awarded the Nobel Prize in economics (officially, the Swedish National Bank's Economic Prize) along with two other game theorists. Nash was the central character of Sylvia Nasar's bestselling Pulitzer Prize nominated biographical book "A Beautiful Mind." A movie based on this book was also popular and received a number of Academy awards in 2002. On May 19, 2015, he received Abel prize from King of Norway. Four days later on his way back home, Nash tragically died when the taxi he was riding crashed with a guard rail on New Jersey turnpike.

Nash's psychiatric diagnosis is well documented but diagnosis of other geniuses suspected of being psychotic is uncertain. However, based on their recorded behaviors, some of them may indeed have had a psychotic condition. One of them is the prominent Spanish surrealistic painter Salvador Dali (Salvador Domingo Felipe Jacinto Dalí i Domènech), who is best known for his painting, "The Persistence of Memory" (Fig. 8.3).

There are many accounts of Dali's bizarre and odd behavior. It is claimed that he had delusions and hallucinations that meet today's criteria for diagnosis of psychosis. He had delusional beliefs that people had stolen ideas from his subconscious mind and from his dreams. His bizarre behavior included signing a book lying on a bed with continuous EKG and EEG recordings, addressing himself as third person in TV interviews, and moving around with an anteater as a pet. Like many individuals with a psychiatric condition, he attempted suicide many times. He also had affairs with multiple women.

FIGURE 8.3 Salvador Dali (1904–89) and his famous painting "The Persistence of Memory," suggesting the fluidity of time. *Reprinted from Wikimedia Commons.*

At first look it is difficult to understand how a psychotic individual with disorganized thought process could be creative. But many of these individuals have periods of remission during which they have minimal symptoms and are less disorganized in thought and action. Their creative talent is possibly expressed during those periods. They are probably also creative even while symptomatic, but due to disorganized thoughts, their ideas are not always expressed in coherent and meaningful ways.

So, why do mental conditions make people creative? As mentioned above altered neural networks could be the primary reason. However, it is not known what alterations are responsible for creative thinking. We have poor understanding of neural alterations both in creative people and in people with psychiatric conditions. But we do know that neurotransmission is dysregulated in individuals with psychiatric conditions.[2–4] Because neurotransmitters control cognition and behavior, altered transmission affects both[5–10], leading to psychiatric symptomatology.[3] The alteration is also responsible for making neural connections unusual, which gives these people the ability to make unusual associations and become creative. Most unusual associations are expressed as delusions or hallucinations in psychosis, but are also expressed as novel ideas. As mentioned above, creativity by definition is creation of novel and unusual associations.[11] Even though not all novel ideas are creative all creative ideas are novel. Therefore, if a person creates novel ideas more frequently, he or she has a greater chance of being creative. Individuals with psychiatric conditions do indeed come up with novel and unusual ideas more often.

It is easy to understand why an abnormal brain leads to novel associations. A "normal" brain learns to make associations that are common. For example, if asked to tell what a brick is used for the most common answer would be "for building a wall," even though it can be used as a door stopper, a paperweight, or a weapon. Since most often bricks are used to build walls, we associate them with that function. A creative brain can think of many more uses. This ability to think in novel ways is called divergent thinking. Based on this ability a test to evaluate the creative ability of individuals was created by an American psychologist J.P. Guilford.[12] The test requires an individual to generate within a fixed time as many uses of a given object as possible. Responses are scored based on several attributes like originality (unusual); fluency (total uses); flexibility (number of categories), etc. A higher score on this test suggests greater creative ability. Interestingly, people with psychotic conditions like schizophrenia generally perform poorly on this task[13] but their relatives score higher than comparable healthy volunteers.[14] Similarly, individuals with schizotypal personality disorder who exhibit bizarre behavior and carry a high risk of developing psychosis also score high.[15,16] Even though these findings are questioned by some researchers,[17] evidence of higher creativity continues to be reported in people with a variety of psychiatric disorders such as people with bipolar disorder.[18] Interestingly the relatives of people with this disorder also show higher creativity, probably due to the high probability of developing this condition.[19]

Altered neural connections are only one of the possible reasons why psychiatric illnesses make people creative. There could be other reasons for their enhanced creative abilities such as lack of hemispheric lateralization. Each hemisphere of the brain is specialized to perform certain tasks more efficiently. For example, in right-handed people the right hemisphere is non-dominant and it specializes in artistic functions while the dominant left hemisphere specializes in tasks that require mathematics or logic. This hemispheric specialization is weak in individuals and family members of people with some form (but not all) of psychiatric illness. Because of the lack of specialization these individuals tend to have equal preference for the use of right and left hand to perform tasks like writing, throwing a ball, etc.[20] Interestingly, mixed handedness is also associated with increased creativity and schizotypal personality.[21] Because of weak lateralization in these individuals all tasks are processed in both hemispheres instead of only in the specialized hemisphere, the normal processing strategy.[22] While lateralization helps the brain process information more efficiently (because of specialized knowledge of certain activities), it limits creativity by using only the strategies it has perfected. It does not allow use of novel strategies for processing or solving problems. Thus, if verbal memory is processed in the left frontal cortex because of lateralization, it would not allow the right frontal cortex to process it using a different strategy. If lateralization is weak, a stimulus can be processed in both hemispheres and take advantage of the information and strategies available in both. That is why people with schizotypal personality disorder activate both hemispheres in tasks of divergent thinking[23] and score better than normal control volunteers, despite exhibiting bizarre behavior, unusual thinking, and egocentric behavior. Because of this ability people with psychiatric conditions perform better in many cognitive tasks including tasks of semantic priming as discussed earlier.[24] Since semantic priming requires making logical connections between two words, better performance suggests that as compared to healthy people, these individuals can make better connections between objects and concepts—both usual and unusual.

A weak hemispheric specialization also provides a novel perspective and hence creates novel ideas. This may explain the creative ability of people with psychiatric conditions like schizophrenia and bipolar disorder in which hemispheric specialization is weak. However, not all individuals with a mental condition have weak hemispheric specialization, even though they also have enhanced creative ability. One of these conditions is attention deficit hyperactivity disorder (ADHD).[25,26] In tests of creativity, people with ADHD outperform healthy control volunteers and are more imaginative in creating novel toys.[26] Obviously, weak hemispheric specialization theory does not explain the creative ability of these individuals. It is not known for sure what makes these people creative. There is very little in common in clinical symptomatology or phenomenology of ADHD and conditions like psychosis and bipolar disorder. One can imagine that conditions in which thought process is compromised could potentially allow individuals to establish unusual thoughts and associations but thought process remains relatively unaffected in ADHD.

However, there is one common link between ADHD and other psychiatric conditions. That link is dopamine. In most psychiatric disorders dopamine neurotransmission is dysregulated. Its dysregulation is well documented in schizophrenia[4] and other psychiatric conditions including ADHD.[3] Could this be the reason for enhanced creativity? It is not clear. But investigators at Karolinska Institute in Sweden recently found a link.[27] They studied dopamine neurotransmission in people with schizophrenia and also in creative people who performed very well in divergent thinking tasks. They observed that both groups have similar changes in dopamine neurotransmission. Specifically, creative healthy people and individuals with schizophrenia have reduced density of D_2 dopamine receptors in the thalamus. How reduced receptor density enhances creativity is unclear but it could be due to increased information flow. The thalamus is a structure through which sensory signals are relayed. It works as a filter to pass selected signals for higher order of processing. By reducing the dopamine receptor this filter or gating mechanism becomes weak and allows more information to pass to the higher brain structures. This additional information could make people more creative.

In addition to reduced dopamine receptor density in the thalamus, there is another mechanism that provides additional information to individuals with schizophrenia. This mechanism involves area V3A. As discussed in Chapter 2, Nonconscious memory, and in other publications,[28–32] this area has a gating mechanism that filters the information released nonconsciously during a cognitive task. If this area is damaged, the gating mechanism weakens resulting in release of more information than normally released during a task. The additional information if processed adequately may enhance cognitive ability but if it remains unprocessed because of cognitive limitation it may cause hallucinations, delusions, confusion and disorganized thinking.[28,33] Previously,[28] I discussed the evidence which suggests that this area is damaged in people with schizophrenia and possibly other psychiatric conditions. As explained earlier, a damaged area V3A could enhance creativity but it could also elicit clinical symptoms such as delusions, hallucinations, and confusion if additional information it releases is incompletely processed. It is not known whether area V3A is partially damaged either structurally or functionally in creative people but this possibility needs to be explored.

The availability of additional information may well be the reason people with ADHD are more creative. These individuals have short attention spans and have difficulty inhibiting unwanted responses. How could these symptoms enhance creativity? It is possible that in a healthy brain, inhibition of "unwanted" responses limits options and may limit the ability to innovate and explore alternate solutions. Since people with ADHD do not inhibit those responses, they can explore additional options. Those options could enhance their creative thinking.[34]

The availability of additional information by itself is not enough to make a person creative. It requires processing of information and conceptualization of its significance. Since individuals with ADHD can process additional information they have enhanced creativity. People who have difficulty processing information to create concepts cannot use additional information to enhance creativity. Autism is an example of a condition in which conceptualization is impaired. That is why despite having a "photographic" memory, these people are not known to be creative in other areas. Fig. 8.4 shows the photographic memory of autistic savant Stephen Wilshire. He painted an amazingly detailed view of the New York skyline after flying over the city for only a few minutes. People with autism have an amazing ability to retain details of perceptual information but cannot use this information to create novel concepts. Ironically, retention of an excessive amount of information may be one of the reasons they have limited ability to conceptualize because it is easier to make concepts when the pool of information is small. A large pool makes it difficult to put the right pieces together to create concepts. Additional information therefore enhances creativity only if the brain has the ability to process it.

Creating a concept is easier if it involves fewer pieces of information. For example, as discussed in earlier chapters, if we had to decide the superiority of a vehicle based on data about gas mileage and price, it would easier to do than if we were given 40 different characteristics of each vehicle

FIGURE 8.4 Savant Stephen Wiltshire painted this New York skyline after he was flown over the city for a few minutes. *Courtesy Stephen Wiltshire.*

(relevant or irrelevant). This is what happens in autistic savants. These individuals do not ignore "irrelevant" stimuli. Therefore, they have difficulty processing a large amount of information they have to deal with to "conceptualize" a situation or a problem.

Further, conceptualization involves creation of association between different pieces of information. Thus, a pen is conceptualized by associating it with an instrument of writing. In real life we are confronted with a variety of sensory information each of which needs to be associated with one another to "conceptualize." Association of multiple stimuli with each other could take a long time. Therefore, in situations that require immediate situational awareness, the brain captures the most significant pieces of sensory stimuli and associates them with the most frequently associated concepts. This process ignores many stimuli and information the brain considers "unimportant." Thus, while reading a book, we focus on words and sentences being read and ignore other stimuli that may be present around us. If those stimuli are not ignored, reading and understanding the text would be difficult. This strategy speeds up processing of the task at hand but ignores information that may or may not be relevant.

Thus, the ability to conceptualize information is a key component of creativity. Since conceptualization is better if it involves a limited amount of information, a disproportionally high number of deaf and blind people have creative talents. Because of their handicap, these individuals do not receive as many sensory "distractors" as normal people do. The limited amount of information helps them process information better and become creative. Some famous blind musicians include Helmut Walcha (1907–91), Andre Marchal (1894–1980), Jean Langlais (1907–91), Jean-Pierre Leguay (born 1939), Ravindra Jain (1944–2015), and jazz composer Arthur Tatum (1909–56). Additionally, many successful country blue performers were blind including Blind Lemon Jefferson, Blind Willie Johnson, Sonny Terry, and Blind Willie Dunn. The bestselling singer in the history of classical music Andrea Bocelli (Fig. 8.5) is also blind. Additionally, the devotional poems of a sixteenth century blind poet Surdas (Fig. 8.6) are still extremely popular in India.

FIGURE 8.5 Andrea Bocelli (b 1958). A top-selling singer who is blind. *From Wikimedia Commons.*

FIGURE 8.6 Surdas (1478–1573). Blind poet and singer, still extremely popular in India, over 400 years after death. *From youtube.com.*

What is most surprising is the ability of blind people to make amazingly vibrant and creative paintings. Blind artists include those who lost sight after birth, later in life, and also those who were born blind. They claim to "see" color by touching the canvas with fingers. The paintings of two blind artists, John Bramblitt and Esref Armagan (Fig. 8.7), are examples of their amazing ability to perceive color, shape, and space. While Bramblitt became blind after an epileptic episode, Armagan was congenitally blind. Other blind creative people include well-known contemporary writers and poets like John Milton, Ved Mehta, Helen Keller, and T.V. Raman.

In addition, there are many athletes and talented people with various handicaps. A recent example is Stephen Hawking (Fig. 8.8), a British physicist who is regarded as one of the most influential contemporary thinkers. He was quadriplegic and could not talk or move limbs. His book *A Brief History of Time* is a popular best seller; it is translated into over 35 languages and has sold over 35 million copies. It appears that a handicap offers a way to minimize interference—by not perceiving certain aspect of stimuli, the brain then focuses on a rather small (and probably unusual) set of stimuli for conceptualization, enhancing creativity.

Apparently, people in ancient times "knew" that one can be more creative by limiting sensory stimuli. In one ancient Indian text, the *Bhagwat Gita* (about 4000 BCE), Lord Krishna describes a learned (genius) person as one who has control over sensory organs. This person is able to hide all senses the way "a tortoise hides legs and head under the shell." Similarly, the *Patanjali Yoga Sutra*[35] suggests that a person can have "true knowledge" only when he or she is in a state of *samadhi* (meditation), which is the highest state of consciousness. In this state all senses are dormant and the practitioner does not perceive any sensory stimulus. In this state, Patanjali says, one gets true vision, which probably involves "visualization" of unusual associations, key to creativity. It suggests that one has to be in an altered state of mind to be more creative!

Thus, creativity is expressed both if the brain has limited information to process and also if it is flooded with information. In the latter case, however, it is important for the brain to be able to process additional information. If it

FIGURE 8.7 Paintings of two blind artists John Bramblitt (top) and Esref Armagan (bottom) show their extraordinary ability to perceive color and space by touch. *Courtesy of respective artists.*

FIGURE 8.8 Stephen Hawking (1942–2018). Regarded as one of the most influential contemporary physicists despite being a paraplegic. *Reprinted from Wikimedia Commons.*

is not processed or incompletely processed, additional information is expressed as delusions, hallucinations, and confusion. It is not clear if creative thinking is processed in a separate neural network or if it uses the network we all use for cognitive processing. The limited evidence available suggests that creative thinking probably uses a separate network, sometimes using a different hemisphere. In most cases, unusual associations leading to creative thinking are generated in the right frontal cortex.[36,37] It is therefore not surprising that neuroimaging and lesion studies have associated the right hemisphere with creative thinking.[38–40]

Creativity therefore is an ability controlled by the nonconscious mind. It not only makes life more interesting but also has survival benefits. Because of their creative abilities our forefathers were able to make weapons to hunt and protect themselves from wild animals. On the flip side, creativity can be used to develop weapons of mass destruction and cause the human race to become extinct.

Bibliography

1. Andreasen NC. *The creating brain: the neuroscience of genius*. Dana Press; 2005.
2. Badgaiyan RD. A novel perspective on dopaminergic processing of human addiction. *J Alcohol Drug Depend* 2013;**1**(1). pii: 1000e101.
3. Badgaiyan RD, Sinha S, Sajjad M, Wack DS. Attenuated tonic and enhanced phasic release of dopamine in attention deficit hyperactivity disorder. *PLoS One* 2015;**10**(9):e0137326.
4. Abi-Dargham A. A dual hit model for dopamine in schizophrenia. *Biol Psych* 2017;**81**(1):2–4.
5. Badgaiyan RD, Wack D. Evidence of dopaminergic processing of executive inhibition. *PLoS One* 2011;**6**(12):e28075.
6. Badgaiyan RD. Dopamine is released in the striatum during human emotional processing. *Neuroreport* 2010;**21**:1172–6.
7. Badgaiyan RD, Fischman AJ, Alpert NM. Dopamine release during human emotional processing. *Neuroimage* 2009;**47**(4):2041–5.
8. Badgaiyan RD, Fischman AJ, Alpert NM. Striatal dopamine release in sequential learning. *Neuroimage* 2007;**38**(3):549–56.
9. Badgaiyan RD, Fischman AJ, Alpert NM. Detection of striatal dopamine released during an explicit motor memory task. *J Nucl Med* 2005;**46**(Suppl. 2):213.
10. Badgaiyan RD, Fischman AJ, Alpert NM. Striatal dopamine release during unrewarded motor task in human volunteers. *Neuroreport* 2003;**14**(11):1421–4.
11. Mednick SA. The associative basis of the creative process. *Psychol Rev* 1962;**69**:220–32.
12. Guilford JP. *Intelligence, creativity, and their educational implications*. 1st ed. R. R. Knapp; 1968.
13. Cropley AJ, Sikand JS. Creativity and schizophrenia. *J Consult Clin Psychol* 1973;**40**(3):462–8.
14. Karksson JL. Genetic association of giftedness and creativity with schizophrenia. *Hereditas* 1970;**66**(2):177–82.
15. Eysenck HJ, Furnham A. Personality and the barron-welsh art scale. *Percept Mot Skills* 1993;**76**(3 Pt 1):837–8.

16. Schuldberg D, French C, Stone BL, Heberle J. Creativity and schizotypal traits. Creativity test scores and perceptual aberration, magical ideation, and impulsive nonconformity. *J Nerv Ment Dis* 1988;**176**(11):648–57.
17. Jaracz J, Patrzala A, Rybakowski JK. Creative thinking deficits in patients with schizophrenia: neurocognitive correlates. *J Nerv Ment Dis* 2012;**200**(7):588–93.
18. Santosa CM, Strong CM, Nowakowska C, Wang PW, Rennicke CM, Ketter TA. Enhanced creativity in bipolar disorder patients: a controlled study. *J Affect Disord* 2007;**100** (1–3):31–9.
19. Simeonova DI, Chang KD, Strong C, Ketter TA. Creativity in familial bipolar disorder. *J Psych Res* 2005;**39**(6):623–31.
20. Cannon M, Byrne M, Cassidy B, Larkin C, Horgan R, Sheppard NP, et al. Prevalence and correlates of mixed-handedness in schizophrenia. *Psych Res* 1995;**59**(1–2):119–25.
21. Claridge G, Clark K, Davis C, Mason O. Schizophrenia risk and handedness: a mixed picture. *Laterality* 1998;**3**(3):209–20.
22. Miran M, Miran E. Cerebral asymmetries: neuropsychological measurement and theoretical issues. *Biol Psychol* 1984;**19**(3):295–304.
23. Folley BS, Park S. Verbal creativity and schizotypal personality in relation to prefrontal hemispheric laterality: a behavioral and near-infrared optical imaging study. *Schizophr Res* 2005;**80**(2–3):271–82.
24. Pizzagalli D, Lehmann D, Brugger P. Lateralized direct and indirect semantic priming effects in subjects with paranormal experiences and beliefs. *Psychopathology* 2001;**34** (2):75–80.
25. Abraham A. Madness and creativity-yes, no or maybe? *Front Psychol* 2015;**6**:1055.
26. Abraham A, Windmann S, Siefen R, Daum I, Gunturkun O. Creative thinking in adolescents with attention deficit hyperactivity disorder (ADHD). *Child Neuropsychol* 2006;**12** (2):111–23.
27. de Manzano O, Cervenka S, Karabanov A, Farde L, Ullen F. Thinking outside a less intact box: thalamic dopamine D_2 receptor densities are negatively related to psychometric creativity in healthy individuals. *PLoS One* 2010;**5**(5):e10670.
28. Badgaiyan RD. Nonconscious processing and a novel target for schizophrenia research. *Open J Psych* 2012;**2**(4A). Available from: https://doi.org/10.4236/ojpsych.2012.224047.
29. Badgaiyan RD. Neuroanatomical organization of perceptual memory: an fMRI study of picture priming. *Hum Brain Mapp* 2000;**10**(4):197–203.
30. Schacter DL, Badgaiyan RD. Neuroimaging of priming: new perspectives on implicit and explicit memory. *Curr Direct Psychol Sci* 2001;**10**:1–4.
31. Badgaiyan RD. Conscious awareness of retrieval: an exploration of the cortical connectivity. *Int J Psychophysiol* 2005;**55**(2):257–62.
32. Badgaiyan RD. Cortical activation elicited by unrecognized stimuli. *Behav Brain Funct* 2006;**2**(17):1–5.
33. Badgaiyan RD. Theory of mind and schizophrenia. *Conscious Cogn* 2009;**18**(1):320–2 discussion 323–324.
34. Abraham A. Is there an inverted-u relationship between creativity and psychopathology? *Front Psychol* 2014;**5**:750.
35. Patañjali, Dvivedi MN, Theosophical publication fund [from old catalog]. *The Yoga-sutra of Patanjali.* Tookárám Tátyá; 1890.
36. Seger CA, Desmond JE, Glover GH, Gabrieli JD. Functional magnetic resonance imaging evidence for right-hemisphere involvement in processing unusual semantic relationships. *Neuropsychology* 2000;**14**(3):361–9.

37. Kiefer M, Weisbrod M, Kern I, Maier S, Spitzer M. Right hemisphere activation during indirect semantic priming: evidence from event-related potentials. *Brain Lang* 1998;**64** (3):377–408.
38. Jung-Beeman M, Bowden EM, Haberman J, Frymiare JL, Arambel-Liu S, Greenblatt R, et al. Neural activity when people solve verbal problems with insight. *PLoS Biol* 2004;**2** (4):E97.
39. Miller L, Tipper L. Effect of focal brain lesions on visual problem solving. *Neuropsychologia* 1996;**34**:387–98.
40. Martindale C. Biological bases of creativity. In: Sternberg RJ, editor. *Handbook of creativity*. Cambridge University Press; 1999. p. 137–52.

Chapter 9

Hypnosis

Cognitively hypnosis is an altered state of consciousness in which an individual has highly focused attention that minimizes competing thoughts and sensations. The term hypnosis was coined by the French magnetizer Etienne Cuvillers (1755–1841) in 1820 and popularized by a Scottish surgeon James Braid (1795–1860). Both Cuvillers' and Braid's interest in hypnosis developed because of their shared belief in the concept of "animal magnetism" developed by a German physician named Franz Mesmer (Fig. 9.1). Mesmer advanced the concept of animal magnetism by suggesting that there is a constant flow of energy and information between living and nonliving objects and that diseases are caused when the flow is disrupted. He developed a method to treat patients by "fixing' the disrupted flow.

Because of his unconventional method of practicing medicine, Mesmer had a difficult career. He treated patients individually and in groups by "resetting disrupted animal magnetism." Since this technique was not approved by the French medical community, in 1784 King Louis XVI appointed a commission to investigate Mesmer's technique. The commission did not find evidence to support the validity of animal magnetism and his techniques.

Mesmer's technique of "resetting" the flow of animal magnetism involved pressing the patient's thumb, looking fixedly into the eyes, making passes along the arm, and pressing the patient's hypochondrium. This procedure was similar to the one practiced in India long before Mesmer popularized it in Europe. An account of the procedure practiced in India was documented by a doctor James Esdaile, who was a part of the British Army serving in India. In his book *Mesmerism in India and its Practical Application in Surgery and Medicine*[1] published in 1846 Esdaile described how physicians use hypnosis to perform painless surgeries for arm, breast, or penile amputations; hydrocele; tooth extraction; and scrotal tumors. Hypnosis was also used to treat nonsurgical medical conditions like eye inflammation, convulsions, sciatica, tactile hallucinations, and lumbago.

Because hypnosis originated in ancient India, it is remarkably similar to the yogic concept of consciousness. Patanjali's *Ashtang Yoga* describes nine stages of consciousness. The sixth state is *dharan*, which is described as a stage of consciousness in which a practitioner concentrates on one object so

Neuroscience of the Nonconscious Mind. DOI: https://doi.org/10.1016/B978-0-12-816115-9.00009-7

FIGURE 9.1 Franz Mesmer (1734–1815). Developed the theory of animal magnetism, which was later popularized as Mesmerism. *Reprinted from Wikimedia Commons.*

intensely that he or she ignores other sensations.[2] That is what happens in hypnosis. Hypnosis is an altered state of consciousness in which an individual has highly focused attention on a suggested task.

Unfortunately, the use of hypnosis for medical and surgical procedures stagnated due to its association with mysticism, quackery, and stage entertainment. However, a select group of people continued to use hypnosis for various purpose. Some of its prominent users include Ovan Pavlov, Alfred Binet, Pierre Janet, William James, Earnest Hillgard, Borsi Sidis, and Sigmund Freud. Even though Freud abandoned hypnosis in his clinical practice in later years, he always acknowledged its role in psychoanalysis.

Hypnosis is performed in two stages: induction and suggestion. For induction (also called trance) a person is asked to focus attention on a real (e.g., breathing) or imaginary (e.g., guided imagery) event. Sometimes it includes suggestions like "you will feel that your arms are relaxed as you listen to my voice." Induction is intended to induce hypnosis and take the patient to an altered state of consciousness. At that state, suggestions are given to alter intended perception or behavior. For example, it may be suggested that a part of his or her body will lose sensation upon hearing a click. Immediately after induction volunteers receive task-specific suggestions. Highly suggestible individuals do not need induction. They can directly advance to the suggestion stage, particularly if the hypnotizer is experienced.

Experienced hypnotizers and highly suggestible individuals can even hypnotize themselves and achieve intended results without anyone's help. This phenomenon is called self-hypnosis. Many investigators believe that even hypnosis induced by a hypnotizer (heterohypnosis) is a form of self-hypnosis. It is claimed that people use the instructions of the hypnotizer to instruct themselves to accomplish a task.

Suggestion is the operative part of hypnosis. At this stage the intended effect—either surgical anesthesia or altered behavior—is achieved. The effectiveness of hypnosis depends on its depth, which in turn depends on the suggestibility of the individual. Since suggestibility is an important factor determining the depth of hypnosis, it is customary to evaluate the suggestibility of before using hypnosis for clinical purposes. This involves assessment of the individual's ability to follow a series of suggestions using one of many tools developed for this purpose. These tools are designed for use in group or individual settings. For group settings the Harvard Group Scale is the most popular tool. In this tool, volunteers are scored on three categories of suggestion: idiomotor-idiosensory direct suggestion, idiomotor challenge, and suggestion related to cognitive alteration. In idiomotor-idiosensory direct suggestion participants are asked to feel heaviness or lightness in limbs. In the idiomotor challenge they are required to overcome this feeling. For example, they are asked to lift a hand against heaviness imposed by the suggestion given earlier. The suggestion related to cognitive alteration includes selective amnesia.

Typically, participants are evaluated in all three categories because an individual can be hypnotizable in one category but not another. Interestingly, there is a hierarchal pattern in susceptibility. Most individuals are suggestible only to idiomotor-idiosensory direct category but not the other two. Only about half of participants suggestible in the idiomotor-idiosensory direct category are suggestible in the diomotor challenge and still a smaller number pass cognition-related suggestions. Because of this hierarchical pattern, those who pass cognition-related suggestion will pass the other two categories. Further, suggestibility runs in families because people with a specific genotype are more suggestible than others. Most high suggestible individuals carry the Met/Val variant of the gene Catechol-*O*-methyltransferase (COMT). These genes regulate the enzyme COMT that is needed to degrade neurotransmitters like epinephrine, norepinephrine, and dopamine. People with other COMT gene variants (Met/Met or Val/Val) have low suggestibility.[3] Because suggestibility is influenced by genes, it is inheritable[4] and remains unchanged throughout adult life.[5]

Other than the tests and traits discussed above, there are not many reliable predictors of hypnotizability. Attempts to find association between personality traits and suggestibility have generally failed, but there are a few interesting reports. It appears that hypnotizable people are more empathic[6] and there is an association between creativity and hypnosis.[7] People with creative imagination are more hypnotizable, and under hypnosis most people become creative. It is understandable why imaginative people are more hypnotizable. They can "visualize" suggestions and instructions better, but it is unclear why people are more creative under hypnosis. Could it be due to their ability to focus on a task more intensely? This is a possibility because by focusing attention on the suggested task and remaining undistracted by other sensory stimuli, hypnotized individuals can process information more efficiently, leading to enhanced creativity. However, this is only a possibility. There is no scientific data to explain why hypnosis enhances creativity.

Our understanding of the neural basis of hypnosis is extremely limited because until recently it was not considered a valid scientific phenomenon. Scientific study of hypnosis received attention after the modern neuroimaging techniques were developed. Since hypnosis is one of those conditions that cannot be studied in animal models, it can only be studied in human volunteers using neuroimaging techniques.

Interestingly, neuroimaging study of hypnosis began with skepticism. Investigators were initially skeptical about sustainability of the hypnotized state inside a scanner, particularly in a magnetic resonance imaging (MRI) scanner, which is noisy and makes people claustrophobic. Additionally, it exposes a volunteer to a strong magnetic field. Initially, investigators thought that in such an environment it would not be possible to sustain a hypnotized state. This concern was addressed by David Oakley and colleagues. At Cardiff University[8] they compared the depth of hypnosis achieved by volunteers on and off scanner and found no significant difference. Volunteers did achieve slightly deeper hypnosis outside the scanner but the difference was not significant. They found that the environment inside the MRI scanner did not interfere with either induction or maintenance of the hypnotic state. Even though this finding encouraged investigators to study hypnosis using neuroimaging techniques, there is little reliable and replicable data on the neural processing of hypnosis. These data are somewhat consistent in associating hypnosis with the right hemispheric activity. This association was first established in the 1970s using an electroencephalogram (EEG), which records electrical activity of the brain.[9] The EEG activity recorded during hypnosis shows an unusual pattern. It shifts from the left hemisphere to the right at the onset of hypnosis. So, if a flash of light is presented to the right eye there is increased EEG activity in the left hemisphere. This activity shifts from the left to the right hemisphere during hypnosis. This shift was observed only in highly suggestible individuals who achieved deep hypnosis. The shift was not observed in individuals with low hypnotizability.[10] This shift of activity may indicate right hemispheric processing of hypnosis. However, because a stroke in the right hemisphere does not affect suggestibility,[11] it is unclear whether hypnosis is controlled exclusively by the right hemisphere.

There is a fair amount of consensus on the correlation between depth of hypnosis and the power of fast EEG waves. There are two types of fast waves: beta and gamma. The frequency of beta waves is typically around 20 Hz and that of gamma over 40 Hz. In contrast, slow waves like alpha and delta have frequencies around 7 and 3 Hz, respectively (Fig. 9.2). Since EEG

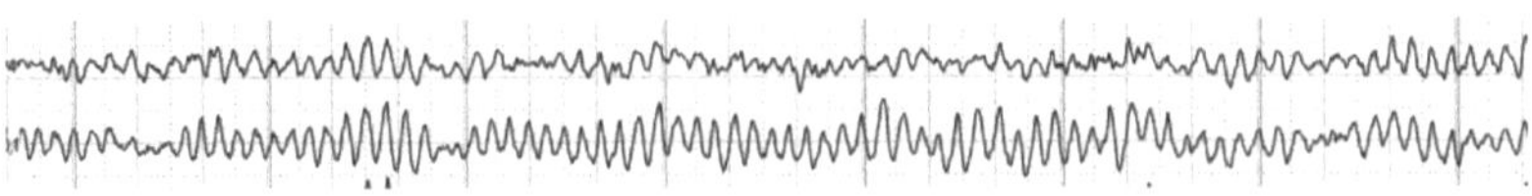

FIGURE 9.2 Examples of the fast (top) and slow (bottom) EEG waves.

is recorded on the scalp surface, it represents the sum of activities of the underlying brain areas. If most neurons in the area covered by an electrode fire (activate) at the same time (synchronized activity), the EEG on the surface will have high amplitude and low frequency (slow wave). Conversely, if the neurons fire at different times, the amplitude of EEG waves will be low and the frequency high (fast wave). The amplitude and frequency of the EEG waves thus indicates degree of synchronization of neuronal activity. Since the depth of hypnosis correlates with fast waves, it is associated with desynchronized neuronal activity.[12]

Another change frequently observed during hypnosis is altered connectivity between brain areas. The alteration was detected by studying the so-called "default mode network."[13,14] This network is active by default when the brain is involved in specific tasks like thinking about self or others, remembering, and future simulations. The brain areas activated during each of these processes are different. The increased or decreased activity observed during these tasks indicates the strength of connection between the areas involved. This network and connectivity is altered in neurological conditions like Alzheimer's disease.[15] A comparison of the activity of the default mode network before and after hypnotic induction reveals reduced activity in the anterior part (primarily frontal cortex) of the brain. Since reduction is observed only in highly suggestible individuals and not in people who do not achieve deep hypnosis, it could be directly associated with hypnosis.[16] While the significance of this reduction is unclear it probably helps reduce attention to stimuli that are not part of the hypnotic suggestion.

In addition, during hypnosis, activity of the attentional network is increased.[17,18] This increase is mostly in the anterior attention network or executive attentional system, which is responsible for selection of targets to focus attention.[19] It indicates that attention is focused on the suggestion during hypnosis. Additionally, there is reduced activity in other frontal areas including the dorsolateral prefrontal cortex.[18] This reduction could be due to suppression of mental activities that are not part of the hypnotic suggestion. This allows volunteers to focus attention on the suggestion without being distracted by other sensory stimuli or spontaneous brain activity. Suppression of the activity of dorsolateral prefrontal cortex by repetitive transcranial magnetic stimulation therefore enhances suggestibility and acceptability of hypnotic suggestions.[20] Suppressed frontal activity is a key neuroimaging feature of hypnosis.

Besides enhancing suggestibility, the dorsolateral prefrontal cortex is also associated with explicit memory and executive function.[21–23] It is not clear how these functions help hypnosis but study of their association with hypnosis will help us better understand the cognitive basis of hypnotic state.

The brain activation induced by hypnosis is unique and distinct from that elicited by mental imagery. Thus, if a volunteer is asked to imagine having pain in the palm, significant activation is observed in the thalamus, but if the

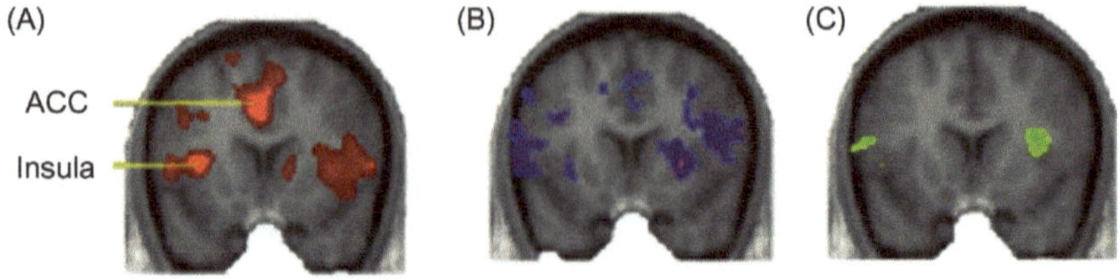

FIGURE 9.3 The brain areas activated during pain induced physically (A); by hypnosis (B); and by imagination (C). *ACC*, anterior cingulate cortex. *Modified from Derbyshire SW, Whalley MG, Stenger VA, Oakley DA. Cerebral activation during hypnotically induced and imagined pain.* Neuroimage *2004;**23**(1):392–401.*

same instruction is given under hypnosis, in addition to the thalamus, a number of other brain areas are activated. These areas include the anterior cingulate cortex, secondary somatosensory cortex, insula, prefrontal cortex, and parietal cortex.[24] This pattern of activation is similar to what we would expect when real pain is inflicted (Fig. 9.3). Hypnosis therefore is not an imagination of the suggested task. It induces physiological changes associated with actual performance of the suggested task. The brain activation induced by hypnotic suggestion and by actual performance of a similar task has also been shown for other tasks.[25,26]

In another interesting neuroimaging study the brain activation elicited during hypnotically induced paralysis of the leg was compared with those elicited by intentionally induced paralysis.[27] As compared to the intentional paralysis, hypnosis elicited greater activation in the orbitofrontal cortex, right cerebellum, left thalamus, and left putamen. This difference suggests that intentional and unintentional activities are processed by separate neural networks. It is an interesting finding that may have forensic significance. It could be used to objectively determine if a person is faking a neurological or perhaps psychiatric symptom. This observation also suggests that under hypnosis people do not intentionally elicit symptoms.

The observation that hypnotically induced changes are involuntary is consistent with the data acquired in the Stroop test.[28] As discussed in Chapter 5 Attention, it is easier to read a word than to name the color of the printed word in the Stroop test. This effect, called the word superiority effect, cannot be voluntarily eliminated because it is an involuntary process controlled by automatic inhibition. This involuntary process can be eliminated if volunteers are given a suggestion during hypnosis that the words they will see will be meaningless symbols and they should not try to attribute any meaning. After this suggestion, if a volunteer is shown a word and asked to name the ink color or read the word, they show no preference for reading. Therefore, the word superiority effect is eliminated.[29,30] Interestingly, this effect can be eliminated in a fully awake nonhypnotized state also by posthypnotic suggestion. Posthypnotic suggestions are given during hypnosis but implemented after the session has ended. For example, during a hypnotic session a volunteer can be instructed that the words he or she will see "after waking up" are useless symbols. This instruction would eliminate the word superiority effect after the session has ended and the volunteer is fully awake.

As mentioned above, the pattern of brain activation elicited during hypnosis is similar to that observed during similar changes induced naturally. However, sometimes the two patterns do not agree because hypnotically induced effects are more intense and severe. This is why hypnosis is sometimes used to enhance the intensity of cognition or to induce involuntary changes in cognition and behavior. It is also used to treat psychosomatic conditions like irritable bowel syndrome,[31] fibromyalgia,[32] and phantom limb.[33] Additionally, hypnosis has been used with limited success to treat psychiatric conditions like obsessive compulsive disorder,[34] delusions,[35] hallucinations,[36] eating disorder,[37] depression,[38] and posttraumatic stress disorder.[39] Because of its ability to alter cognitive processing, hypnosis is an excellent model to study conditions like conversion disorder, hemineglect, and movement disorders.

In recent years hypnosis has gained acceptance as a psychosomatic phenomenon that can modulate cognition and behavior. With this acceptance, hopefully it will be used more often as a therapeutic and diagnostic tool. It could be an important technique to study the neural basis of conscious and nonconscious experience—the experience that makes us human.

Bibliography

1. Esdaile J. *Mesmerism in India and its practical application in surgery and medicine.* Longman, Brown, Green, and Longmans; 1846.
2. Mehta J. *Essence of Maharishi Patanjali's Ashtang Yoga.* Pustak Mahal; 2005.
3. Lichtenberg P, Bachner-Melman R, Gritsenko I, Ebstein RP. Exploratory association study between Catechol-O-Methyltransferase (COMT) high/low enzyme activity polymorphism and hypnotizability. *Am J Med Genet* 2000;**96**(6):771–4.
4. Morgan AH. The heritability of hypnotic susceptibility in twins. *J Abnorm Psychol* 1973;**82**(1):55–61.
5. Piccione C, Hilgard ER, Zimbardo PG. On the degree of stability of measured hypnotizability over a 25-year period. *J Pers Soc Psychol* 1989;**56**(2):289–95.
6. Wickramasekera II IE, Szlyk JP. Could empathy be a predictor of hypnotic ability? *Int J Clin Exp Hypnosis* 2003;**51**(4):390–9.
7. McHenry RE, Shouksmith GA. Creativity, visual imagination and suggestibility: their relationship in a group of 10-year-old children. *Brit J Educat Psychol* 1970;**40**(2):154–60.
8. Oakley DA, Deeley Q, Halligan PW. Hypnotic depth and response to suggestion under standardized conditions and during fMRI scanning. *Int J Clin Exp Hypnosis* 2007;**55**(1):32–58.
9. Frumkin LR, Ripley HS, Cox GB. Changes in cerebral hemispheric lateralization with hypnosis. *Biol Psych* 1978;**13**(6):741–50.
10. Naish PL. Hypnosis and hemispheric asymmetry. *Conscious Cogn* 2010;**19**(1):230–4.
11. Kihlstrom JF, Glisky ML, McGovern S, Rapcsak SZ, Mennemeier MS. Hypnosis in the right hemisphere. *Cortex* 2013;**49**(2):393–9.

12. Cardena E, Jonsson P, Terhune DB, Marcusson-Clavertz D. The neurophenomenology of neutral hypnosis. *Cortex* 2013;**49**(2):375–85.
13. Gusnard DA, Raichle ME, Raichle ME. Searching for a baseline: functional imaging and the resting human brain. *Nat Rev Neurosci* 2001;**2**(10):685–94.
14. Raichle ME, MacLeod AM, Snyder AZ, Powers WJ, Gusnard DA, Shulman GL. A default mode of brain function. *Proc Natl Acad Sci USA* 2001;**98**(2):676–82.
15. Greicius MD, Srivastava G, Reiss AL, Menon V. Default-mode network activity distinguishes Alzheimer's disease from healthy aging: evidence from functional MRI. *Proc Natl Acad Sci USA* 2004;**101**(13):4637–42.
16. McGeown WJ, Mazzoni G, Venneri A, Kirsch I. Hypnotic induction decreases anterior default mode activity. *Conscious Cogn* 2009;**18**(4):848–55.
17. Deeley Q, Oakley DA, Toone B, Giampietro V, Brammer MJ, Williams SC, et al. Modulating the default mode network using hypnosis. *Int J Clin Exp Hypnosis* 2012;**60**(2):206–28.
18. Jiang H, White MP, Greicius MD, Waelde LC, Spiegel D. Brain activity and functional connectivity associated with hypnosis. *Cereb Cortex* 2017;**27**(8):4083–93.
19. Posner M, Badgaiyan R. Attention and neural networks. In: Parks RW, Levine DS, editors. *Fundamentals of neural network modeling: neuropsychology and cognitive neuroscience.* The MIT Press; 1998. p. 61–76.
20. Dienes Z, Hutton S. Understanding hypnosis metacognitively: rTMS applied to left DLPFC increases hypnotic suggestibility. *Cortex* 2013;**49**(2):386–92.
21. Badgaiyan RD, Posner MI. Time course of cortical activations in implicit and explicit recall. *J Neurosci* 1997;**17**(12):4904–13.
22. Badgaiyan RD. Neuroanatomical organization of perceptual memory: an fMRI study of picture priming. *Hum Brain Mapp* 2000;**10**(4):197–203.
23. Badgaiyan RD. Executive control, willed actions, and nonconscious processing. *Hum Brain Mapp* 2000;**9**(1):38–41.
24. Derbyshire SW, Whalley MG, Stenger VA, Oakley DA. Cerebral activation during hypnotically induced and imagined pain. *Neuroimage* 2004;**23**(1):392–401.
25. McGeown WJ, Venneri A, Kirsch I, Nocetti L, Roberts K, Foan L, et al. Suggested visual hallucination without hypnosis enhances activity in visual areas of the brain. *Conscious Cogn* 2012;**21**(1):100–16.
26. Mendelsohn A, Chalamish Y, Solomonovich A, Dudai Y. Mesmerizing memories: brain substrates of episodic memory suppression in posthypnotic amnesia. *Neuron* 2008;**57**(1):159–70.
27. Ward NS, Oakley DA, Frackowiak RS, Halligan PW. Differential brain activations during intentionally simulated and subjectively experienced paralysis. *Cognit Neuropsych* 2003;**8**(4):295–312.
28. Stroop JR. Studies of interference in serial verbal reactions. *J Exp Psychol* 1935;**18**:643–62.
29. Zahedi A, Stuermer B, Hatami J, Rostami R, Sommer W. Eliminating Stroop effects with post-hypnotic instructions: brain mechanisms inferred from EEG. *Neuropsychologia* 2017;**96**:70–7.
30. Raz A, Shapiro T, Fan J, Posner MI. Hypnotic suggestion and the modulation of Stroop interference. *Arch Gen Psych* 2002;**59**(12):1155–61.
31. Palsson OS. Standardized hypnosis treatment for irritable bowel syndrome: the North Carolina protocol. *Int J Clin Exp Hypnosis* 2006;**54**(1):51–64.

32. Picard P, Jusseaume C, Boutet M, Duale C, Mulliez A, Aublet-Cuvellier B. Hypnosis for management of fibromyalgia. *Int J Clin Exp Hypnosis* 2013;**61**(1):111–23.
33. Oakley DA, Whitman LG, Halligan PW. Hypnotic imagery as a treatment for phantom limb pain: two case reports and a review. *Clin Rehabil* 2002;**16**(4):368–77.
34. Meyerson J, Konichezky A. Hypnotically induced dissociation (HID) as a strategic intervention for enhancing OCD treatment. *Am J Clin Hypnosis* 2011;**53**(3):169–81.
35. Attewell JE, Cox RE, Barnier AJ, Langdon R. Modeling erotomania delusion in the laboratory with hypnosis. *Int J Clin Exp Hypnosis* 2012;**60**(1):1–30.
36. Ortega DF. Hypnosis in the treatment of hypnopompic hallucinations: a case report. *Am J Clin Hypnosis* 1984;**27**(2):111–13.
37. Barabasz M. Efficacy of hypnotherapy in the treatment of eating disorders. *Int J Clin Exp Hypnosis* 2007;**55**(3):318–35.
38. Alladin A, Alibhai A. Cognitive hypnotherapy for depression: an empirical investigation. *Int J Clin Exp Hypnosis* 2007;**55**(2):147–66.
39. Lynn SJ, Cardena E. Hypnosis and the treatment of posttraumatic conditions: an evidence-based approach. *Int J Clin Exp Hypnosis* 2007;**55**(2):167–88.

Chapter 10

Extrasensory perception

Traditionally, the knowledge we gain without involving known sensory system is collectively called extrasensory perception, or simply ESP. It includes phenomenon like telepathy, clairvoyance, precognition, and retrocognition.

The knowledge of, and fascination with, ESP is probably as ancient as human civilization. Its earliest and most vivid account is found in the Indian scripture *Bhagwat Gita*, scripted around 500 BCE but existed much earlier. It describes in detail how a charioteer of King Dhritarashtra named Sanjay could “see” in real time the war being fought between the king’s sons (kauravas) and nephews (pandavas) over 100 miles away. Sanjay gave the king not only running commentary of the battlefield but also relayed what lord Krishna was preaching to one of the warriors Arjun in the battlefield. He could therefore not only “see” but also “listen” to what was happening miles away. This mythological story does not necessarily mean that Sanjay had extraordinary power, but it does suggest that people in ancient times were familiar with the concept of distant vision and ESP.

Other forms of ESP have also been described in ancient literature. It is therefore surprising that we cannot confirm or deny existence of a phenomenon that has been known for thousands of years. There are many reasons we do not know much about ESP but plurality of terminology and lack of an acceptable definition are among possible reasons.

Since there is no accepted definition, for the purpose of this chapter we define ESP as a phenomenon that allows acquisition of information without using technology or known sensory modalities (sight, hearing, smell, taste, touch, proprioception, etc.). We think this “definition” clarifies some of the controversies concerning this phenomenon. It is often incorrectly defined as the knowledge acquired without using a sensory receptor. Since perception without a receptor is not consistent with the current scientific thinking, many scientists stayed away from ESP research – just like they stayed away from studying consciousness not too long ago.

I think it is inaccurate to characterize ESP as perception without a sensory receptor. If it really exists it must have receptors. However, it is possible that those receptors have not yet been discovered or identified. If we accept this assumption, it may be relatively easy to study the science of ESP

Neuroscience of the Nonconscious Mind. DOI: https://doi.org/10.1016/B978-0-12-816115-9.00010-3

and advance our knowledge not only of this phenomenon but also of other biological processes. I think no biological phenomenon should be out of bounds for science—definitely not a phenomenon that has been known for centuries.

There are several reasons we do not understand biological processes like ESP. One of the reasons is the use of laws borrowed from physics and chemistry to understand these processes. Sometimes we forget that the laws of physics and chemistry are formulated to explain the behavior of nonliving objects. Therefore, these laws are not expected to explain processes that are uniquely biological. Unless theories that explain biological processes are developed, we will not be able to fully understand uniquely biological phenomena like ESP and consciousness. Ironically, to develop those theories we need to study phenomena that are unique to life—the phenomena we do not understand!

In this chapter I discuss some of the findings and emphasize the need for further systematic scientific study of this phenomenon. I believe we should continue to study these processes until their neural bases are understood or until it is proved that these phenomena do not exist.

To be fair, quite a few researchers made serious attempts to understand ESP in the past but at the time they had limited tools to study the human brain. That is why after numerous studies conducted in the last century, enthusiasm reduced significantly. These studies did not prove conclusively whether ESP exists or not. Let us examine what we know about the science of ESP.

The interest of the general public and scientists in ESP peaked in the early- and mid twentieth century primarily due to people practicing witchcraft. They claimed to have magical powers including the ability to "see" the future and control people's mind. These individuals were feared because of their purported abilities, and this fear was so intense that the British government outlawed witchcraft and arrested individuals practicing it. Despite the law, people believed in their extraordinary abilities. Even the British prime minister Winston Churchill believed them. It is widely reported that because of this belief he helped a British medium Helen Duncan get out of prison. She was arrested for practicing witchcraft in 1944. In a séance she claimed that a British sailor had contacted her to say that his ship had just sank and everyone onboard had died. The sailor wanted to talk to his mother who happened to be in the séance at the time. Since Britain had not publicly acknowledged sinking of the ship, Ms. Duncan was arrested on the charge of practicing witchcraft. Winston Churchill believed in her telepathic ability and helped her get out of prison. There is no dearth of similar stories around the world. Some of them - probably most of them - are just stories but quite a few may be real. It is however, a common knowledge that at times we do get premonition about an event that would happen in future. We don't know for sure if that happens because of our conscious or nonconscious knowledge of factors that lead to those events or because of ESP abilities.

We are not even close to answer this question yet because systematic scientific study of ESP began only about a hundred years ago. The first academic program to study the phenomenon started in 1911 at Stanford University using funds donated by Thomas Stanford, the brother of the university's founder. But ESP research really took off when an American botanist Joseph Banks Rhine (Fig. 10.1), popularly known as J.B. Rhine, defined its scope, designed scientific experiments, and acquired data to demonstrate the ESP effect. Rhine became interested in ESP as a graduate student in Chicago after attending a lecture by the famous physician and novelist Sir Arthur Conan Doyle who created the fictional detective character Sherlock Holmes. Doyle talked about "scientific evidence" of communication with dead people. Rhine was fascinated and after finishing Ph.D. in botany moved to Harvard University to study psychology. Later, he accepted a faculty position at Duke University and setup a laboratory to study ESP. Duke was the second major university to establish ESP program after Stanford.

At Duke, Rhine developed several experimental protocols and with the help of a colleague Karl Zener designed a special set of cards, called Zener cards, to use in ESP experiments. Each card has one of five symbols on it. In a typical ESP experiment, after hiding the face of the card, volunteers are asked to guess the symbol on it. One of Rhine's volunteers, Adam Linzmeyer, reportedly responded correctly 100% of the time in a 9-card deck; 36% of the time in a 25-card deck; and 40% of the time in a 300-card deck. Another volunteer named Hubert Pearce (Fig. 10.1) consistently scored 40% in 25-card tests. The random chance of making correct response in these cards is 20%. Rhine also developed double-blinded experiments to objectively quantify the ESP effect. He documented his experiments in a

FIGURE 10.1 Hubert Pearce (left) with J.B. Rhine (1885–1980) who developed experimental paradigms to study ESP. *Reproduced from Wikimedia Commons.*

book called *New Frontiers of the Mind: The Story of Duke Experiments.*[1] Along with the chief of the ESP unit William McDougal, he coined the term parapsychology and founded the Institute for Parapsychology at Duke It is no longer affiliated to the university and has now been renamed Rhine Research Center. He also founded the Parapsychological Association and published the *Journal of Parapsychology*.

Rhine reported significant ESP effects in the experiments he conducted at Duke. Since his data were impressive, other researchers tried to replicate. None of his data were reliably replicated.[2–4] In 1936 a large study was conducted at Princeton University to find evidence of ESP using Rhine's method. The study was led by W.S. Cox who used playing cards to study ESP in 132 volunteers. He gave them over 25,000 trials but did not find any evidence of ESP.[5] When doubts were raised about possible flaws in methodology, Rhine disclosed in the book, *Extra Sensory Perception After Sixty Years*,[6] that one of his volunteers repeatedly cheated during experiments. It was also discovered that the symbols printed on the cards allowed volunteers to see from the back of the cards. Rhine thus stopped using cards and started using dice. In these experiments he claimed volunteers guessed above-chance level the face on a dice before it was rolled. His data were impressive but not everybody believed that above-chance performance was due to ESP, because dices have a tendency to produce a specific face more often than others because of its design. This tendency could be picked up by volunteers consciously or nonconsciously to give above-chance level of prediction.

For these reasons the reliability of Rhine's data was questioned. Additionally, he was also alleged to have manipulated statistical methods to make the effect appear more significant than it actually was. But the inability of other investigators to replicate Rhine's findings was the most serious problem. Incidentally, Rhine was not the only researcher whose ESP experiments were not replicated. Experiments conducted by other investigators also had the same problem. In some cases researchers were unable to replicate results of their own experiments. For example, a British psychologist named Keith M. Hearne reported observing visual evoked potentials in receiver after these potentials were generated in another person by showing visual stimuli.[7] Since other laboratories were unable to replicate his "findings," Hearne decided to repeat the experiments himself but was unable to replicate his previous results. He published the results of the second experiment four years later.[8]

Despite these controversies interest in ESP continued unabated. But investigators this time were more careful. They designed rigorous protocols to make experiments replicable. One of those protocols is called the ganzfeld procedure (German word meaning "entire field") procedure, a modification of the protocol developed by German psychologist Wolfgang Metzer to study the visual field. In this procedure an environment of sensory deprivation is created. Assuming that this environment would enhance ESP and make the results replicable, the procedure was used to study ESP. The ganzfeld procedure was first used in an ESP experiment in the 1970s by an American parapsychologist Charles Honorton (Fig. 10.2). He studied telepathy at the Maimonides Medical Center in New York. True to his assumption, data were much cleaner and he observed strong telepathic signals.[9] This experiment significantly influenced ESP research and the ganzfeld procedure remains today the mainstay of parapsychology research.

In the ganzfeld procedure a volunteer who is called a receiver is placed comfortably in a room filled with red light. The receiver is blinded by closing eyes with a ping pong ball cut into two halves. A second person called the sender located in another room selects one of the four targets and tries to mentally communicate his or her selection to the receiver. The experimenter who remains blinded to the selection, records the receiver's response. Later, the receiver is taken out of the ganzfeld and asked to select one of the four targets that resembles the object selected by the sender. Surprisingly, in this study receivers select the correct target at a significantly greater rate than expected by chance.[10] This finding was replicated by many investigators in different laboratories.

It is claimed that using a similar technique a sender can even alter a receiver's physiological parameters. In one such experiment conducted at the Mind Science Foundation in San Antonio Texas, William Braud and Marilyn Schlitz showed the sender (also called the influencer) a video of the receiver sitting or meditating in another room. Now and then the influencer would stare at the receiver's video. They found that the stare would make the

FIGURE 10.2 Charles Honorton (1946–92). Popularized ganzfeld protocol for use in ESP experiments. *Image courtesy Parapsychological Association.*

receiver anxious and alter his or her galvanic skin response (GSR) indicating anxiety.[11] The receiver had no clue about the time of staring and the reason for altered GSR. This highly influential experiment was well designed and investigators conducted 13 experiments over 13 years using dozens of volunteers. The results reliably demonstrated the ESP effect.

In another influential study T.D. Duane and Thomas Behrendt of Jefferson Medical College Philadelphia studied 15 pairs of identical twins.[12] They placed each twin in a separate room and recorded their electroencephalogram (EEG). The experimenter then induced changes in the EEG of one of the twins by asking him or her to open or close eyes in a lighted room. When the eyes are open, high-frequency beta waves dominate the EEG. These waves convert to low-frequency alpha waves when the eyes are closed. Duane and Behrendt found that induction of alpha waves in one twin induced a similar change in the other who was unaware of these changes. However, this effect was observed only in 2 of the 15 pairs of twins studied. It is not clear why the other 13 pairs did not show this effect. The two pairs who showed correlated EEG activity participated in another experiment conducted by Leanne Standish in Washington University. In this study they were separated by electrical and magnetic shielding. A light was shown to one of them while EEG and fMRI signals of the other were recorded. Surprisingly, both the EEG and fMRI signals consistent with flashing light were observed in the nonstimulated member of the pair.[13,14] As in the EEG experiment, this effect also could not be elicited in all twins. It is estimated that 15%–20% twins show this telepathic ability.

Over the years, people have struggled to understand why only a few people and a small number of twins show the ESP effect. Skeptics suggest that the positive results in those few people were due to the use of faulty statistics and flawed experimental procedures. To address these issues and to make experiments replicable, Charles Honorton collaborated in 1986 with an ESP skeptic Ray Hyman, Professor of Psychology at the University of Oregon, to jointly design a "flawless" experiment and statistical procedure.[15] The experiment Honorton had conducted earlier using the ganzfeld procedure was repeated under a revised protocol. A receiver was asked to guess one of the four target cards that a sender sitting in another room selected. The revised protocol produced a correct result 32% of the time, which was significantly higher than the 25% expected by chance. However, it was discovered that greater-than-chance performance depends on the frequency of target presentation. If the target was presented only once, performance was at chance level but for targets that were presented six or more times, the correct response was 52%. It did not make much sense but prompted people to question the validity of statistical methods used in ESP experiments.

Publication of this data initiated debate between people who believed in ESP and those who did not. It led the US National Academy of Science in 1988 and the US Congressional Office of Technology Assessment in 1989 to investigate the validity of ESP data. Unfortunately, those investigations did not arrive at any conclusion. Skeptics and believers held their respective positions and debate on the validity of ESP experiments continued unabated. However, those who believed had the support of the Defense Intelligence Agency of the US Army. In 1978 it started a 20-million-dollar project on ESP, codenamed Stargate. The objective of this project was to examine whether ESP could be used in future warfare. The project was later transferred to the Central Intelligence Agency (CIA) and officially terminated in 1995 but unofficially continued much longer (perhaps even today) at a lower level of funding. The data acquired and experiments conducted in the project were classified until recently. In January of 2017 the CIA declassified data and made it freely available on their website.

Stargate had 15–20 staff stationed at Fort Meade Maryland, close to Washington DC. It was directed by a nuclear physicist turned parapsychologist Edwin C. May. He recently coedited a book called *Extrasensory Perception: Support, Skepticism and Science*[16] where he critically reviewed the ESP data acquired in Stargate and elsewhere. He claims that the project had made significant progress in distant vision. As an example, May cites a remote viewer named McMoneagle who was able to remotely "see" a new class of Soviet submarine being built. He also accurately predicted the date of its launch. The submarine was a Typhoon class type and was launched around the time the remote viewer predicted. It was remarkable telepathic information. However, veracity of this and other claims has never been established because May used to closely guard the data and did not allow independent verification. His claims about successful ESP experiments have thus never been endorsed either by the scientific community or by government committees with oversight authority. No significant discoveries were made and publically available information suggests that none of its "findings" were used by the military or any other government agency. Even though the project did not prove existence or usefulness of ESP beyond a reasonable doubt, it did provide legitimacy to ESP research.

This legitimacy encouraged investigators to look at ESP more seriously. It was soon recognized that besides questionable experimental design and statistical method there was another issue with these experiments— lack of data on negative results. Since experiments with negative outcomes are generally not reported, it was difficult to estimate ratio of successful versus unsuccessful ESP experiments. Negative results are needed to determine whether positive data were acquired by chance. If, for example, only 1 out of 100 experiments found an ESP effect that will almost certainly be considered an outlier observed by chance. It cannot be used to validate ESP. Similarly, it is not known how many trials are needed to arrive at a conclusion. Thus, if we flip a coin there is a 50–50 chance of getting a head or tail assuming that the coin has no imperfection. However, if we take a small sample we may not get a 50–50 split. If we flip it six times, it is quite possible that we will get four heads and two tails by chance. In an experiment a similar situation could falsely lead us to conclude significant bias for the head to show up. This "bias" could be reversed in the next six flips. Could similar biases be responsible for "successful" ESP experiments? This possibility cannot be discounted particularly because most of these experiments were not reliably replicated.

But to be fair, based only on the above argument, the ESP effect cannot be completely ruled out because a number of factors affect the outcome of an experiment on human cognition and behavior. These factors include the mood and mental state of the volunteer and experimenter, time of the day, milieu of the laboratory, and many other factors that we may not be even aware of such as geomagnetism, season of the year, etc. Incidentally, there is evidence of significant seasonal variation in ESP ability.[16] Since we do not know all the factors that affect an ESP experiment, it is impossible to control those factors and replicate findings of another experiment. In addition, like hypnosis, some volunteers are likely to have better ESP ability than others. It is thus important that initial ESP experiments be conducted only on people with known ESP ability.

In recent years investigators have used neuroimaging methods to address some of the controversy concerning the validity of ESP experiments. Initial studies used EEG to look for changes in the brain activity while volunteers performed an ESP task. Most of these studies reported abnormally high density of slow alpha EEG waves. As discussed in Chapter 9, Hypnosis, slow waves indicate greater synchronization of neuronal firing. In an experiment, Rex G. Stanford of St. John's University had volunteers listen to Indian flute music for about 8 minutes and free associate with a hidden picture selected randomly from a magazine. They were asked to guess contents of the picture while listening to the music. Some of them made more accurate guesses than others. Based on these guesses, volunteers were divided into two groups: Those who guessed at significantly above-chance level ($P < .2$) and those who guessed at chance. EEGs recorded during free association revealed high density of alpha waves (37.54%) in volunteers who performed above chance as compared to those who performed at chance level (12.13%). It was most significant in the occipital region.[17]

Since increased alpha density did not provide much information about the neural processing of ESP except suggesting increased synchronization, a Harvard psychologist Stephen Kosslyn used fMRI to study brain activity.[18] In the scanner volunteers were asked to look at pictures designated either as ESP or non-ESP. The ESP pictures were either shown simultaneously to another person at a separate location to elicit telepathy or presented afterward to evoke precognition. Non-ESP pictures were used as control. In the experiment ESP and non-ESP pictures activated the same brain areas. Based on this observation Kosslyn concluded that ESP does not exist because stimuli associated with ESP were not processed differently. The results however, were not considered reliable because of several problems with the design of this experiment. The experiment did not take into account the fact that not everybody has ESP ability and it is not known whether ESP effect was elicited in the experiment. Moreover, mixing different kinds of ESP tasks in one experiment is also not a good idea in an experiment of this kind. Because of these and other issues, results of this experiment are not considered credible.

Another study published in the same year (2008) had a better design and it arrived at a different conclusion. This study was conducted by Venkatasubramanian and colleagues[19] at the National Institute of Mental Health and Neuroscience, Bengaluru, India. They used fMRI to scan a control volunteer and a "mentalist" Mr. Gerard Senehi who is known to possess telepathic ability. The mentalist and control volunteer were scanned while they tried to guess a figure being drawn by the experimenter located in another room. The mentalist surprisingly made almost perfect guess while the control volunteer did not (Figs. 10.3 and 10.4). There was therefore evidence of telepathy in this experiment. The pattern of brain activity observed in the two volunteers was significantly different. The mentalist had increased activation in the right parahippocampal gyrus while the control volunteer had increased activation in the left inferior frontal gyrus. It is not clear how the parahippocampus gyrus helps ESP, but it is obvious that different neural networks are involved in ESP and guesswork without ESP. Even though both volunteers tried to "guess" the same object, different areas of their brains were activated based on whether ESP was evoked or not.

In another similar experiment conducted in Hawaii, a group of "healers" who claimed to have the ability to telepathically connect with other people were asked to communicate with a receiver who was being scanned using

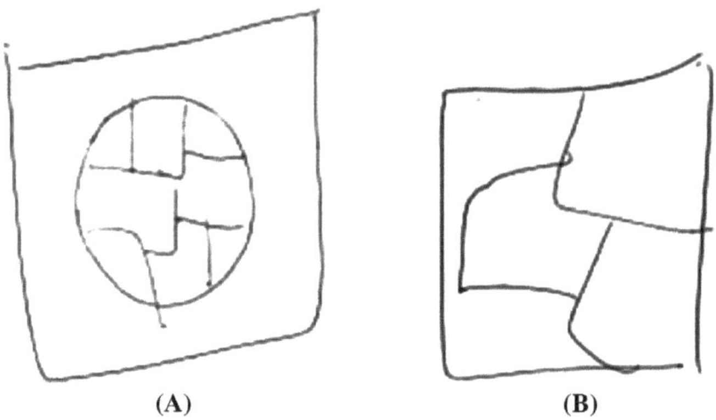

FIGURE 10.3 Pictures drawn by the experimenter (A) and a mentalist (B). *Reproduced from Venkatasubramanian G, Jayakumar PN, Nagendra HR, Nagaraja D, Deeptha R, Gangadhar BN. Investigating paranormal phenomena: functional brain imaging of telepathy.* Int J Yoga *2008;**1**(2):66–71.*

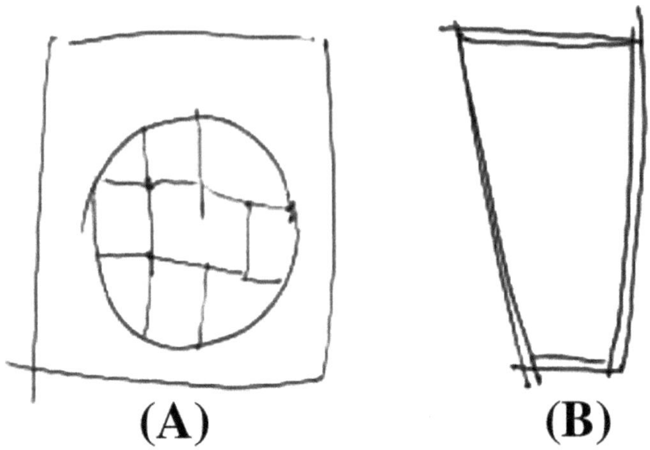

FIGURE 10.4 Picture drawn by the experimenter (A) and a control volunteer (B). *Reproduced from Venkatasubramanian G, Jayakumar PN, Nagendra HR, Nagaraja D, Deeptha R, Gangadhar BN. Investigating paranormal phenomena: functional brain imaging of telepathy.* Int J Yoga *2008;**1**(2):66–71.*

fMRI. When healers sent telepathic signals, the receiver showed increased activation in the anterior cingulate cortex, superior frontal gyrus and precuneus, close to the occipito-parieto-temporal (OPT) junction.[20] Another person with known ESP ability, Sean Harribance of Trinidad, was studied by W.G. Rolls and colleagues at the University of West Georgia. Mr. Harribance reportedly developed his ability to identify hidden objects after he suffered a head injury at the age of 13. In an experiment he was asked to describe photographs hidden inside opaque envelopes. While he was telepathically trying to "see" pictures, his EEG activity increased in the right OPT region. The metabolic activity recorded using single photon emission computed tomography (SPECT) also increased in the same area.[21] In yet another experiment a Canadian group used EEG and fMRI to study Ingo Swann who is known to have telepathic ability.[22] Mr. Swann was positioned in the scanner and asked to draw or describe a distant location that he had never visited. He showed telepathic ability by correctly drawing those locations in most trials. In this remote viewing task, there was increased density of alpha waves in EEG. Analysis of the fMRI data revealed abnormal activity in the right OPT junction.

The findings of high density of alpha waves and activation in the OPT junction across different laboratories and different volunteers is remarkable. It underscores the importance of OPT junction in the processing at least in people with known ESP abilities. This finding is interesting because in a series of neuroimaging experiments we have shown that this area is involved in retrieval of nonconscious information.[23–33] As discussed in Chapter 2, Nonconscious memory, area V3A located at the OPT junction is involved in the retrieval of information needed to perform a task. Further, the retrieved information remains nonconscious until it is processed in the V3Afrontal loop. This processing makes us consciously aware of the information. Thus, involvement of this area in ESP experiments helps us characterize ESP as a nonconscious process that uses the same neural network that processes other nonconscious functions. But it is not clear why only some of us are able to use this network to elicit ESP. Could it be due to individual variation in abilities? Possibly yes. It is not uncommon for some of us to possess physical and mental abilities that others do not. We are all unique with individual talents, viewpoints, likes, dislikes, habits, etc. Because of this uniqueness some of us are better at predicting the future and make better long-term decisions while others make decisions that are detrimental to their future prospects.

Most of us at some point in life experience dreams or premonitions that come true. We do not know why this happens occasionally. Could it be due to certain geomagnetic phenomenon or because we are exposed to some kind of undiscovered stimulus or influence? It is also possible that when conditions are right, our brain changes its state and opens up ESP abilities for us. This possibility is philosophically described in ancient literature including the *Patanjali Yoga Sutra*. These scriptures suggest that people acquire extraordinary abilities including ESP by practicing meditation. According to these texts if external sensory organs are suppressed, the brain takes us to another state of consciousness. In that state the brain acquires extraordinary ability to perceive and understand signals that are not ordinarily perceived and understood.[34,35]

Scientifically, we know we have many "hidden" sensors in addition to the five senses (vision, hearing, taste, touch, and smell) that we are aware of. For example, we have sensors that detect blood pressure (baroreceptors), chemical composition of body fluids (chemoreceptors), position of our joints (proprioceptors), etc. These sensors detect a variety of stimuli but we remain unaware because their activities remain nonconscious. Is it possible that we have similar hidden sensors for ESP? It is not beyond the realms of possibility. It is possible that we have not yet discovered those sensors and do not know under what conditions they get activated.

However, we do know that some of the hidden sensors can be voluntarily controlled by practice. People are known to alter their heart rate and even brain activity at will[36,37] and under hypnosis one can increase or decrease blood flow and skin temperature.[38] These activities usually remain out of voluntary control, but it is not impossible to gain at least limited control over them. It is therefore not surprising that some of us have access to ESP function while most of us do not. If ESP is really mediated through the hidden sensory system, the next logical question is where are these sensors located and what kind of stimuli do they respond to? Obviously, we do not know the answers. But it provides a road map for investigators to explore. I have discussed these issues further in Chapter 13 Future outlook.

Bibliography

1. Rhine JB. *New frontiers of the mind; the story of the Duke experiments.* Farrar & Rinehart; 1937.
2. Lawson TJ. *Scientific perspectives on pseudoscience and the paranormal: readings for general psychology.* Pearson Prentice Hall; 2007.
3. Hines T. *Pseudoscience and the paranormal.* 2nd ed. Prometheus Books; 2003.
4. Smith JC. *Critical thinking: pseudoscience and the paranormal.* 2nd ed. Wiley; 2017.
5. Cox JW. An experiment in ESP. *J Exp Psychol* 1936;**12**:437.
6. Pratt JG, Rhine JB. *Extra-sensory perception after sixty years: a critical appraisal of the research in extra-sensory perception.* Bruce Humphries; 1967.
7. Hearne K. Visually evoked responses and ESP. *J Soc Psych Res* 1977;**49**:648–57.
8. Hearne K. Visually evoked responses and ESP: failure to replicate previous findings. *J Soc Psych Res* 1981;**51**:145–7.
9. Honorton C, Harper JW. Psi-mediated imagery and ideation. *J Am Soc Psych Res* 1974;**68**:156–68.
10. Storm L, Ertel S. Does psi exist? Comments on Milton and Wiseman's (1999) meta-analysis of ganzfeld research. *Psychol Bull* 2001;**127**(3):424–33 discussion 434–8.
11. Braud WG, Schilitz MJ. Psychokinetic influence on electrodermal activity. *J Parapsychol* 1983;**47**:95–119.
12. Duane TD, Behrendt T. Extrasensory electroencephalographic induction between identical twins. *Science* 1965;**150**(3694):367.
13. Standish LJ, Johnson LC, Kozak L, Richards T. Evidence of correlated functional magnetic resonance imaging signals between distant human brains. *Altern Ther Health Med* 2003;**9** (1):128 122–5.
14. Richards TL, Kozak L, Johnson LC, Standish LJ. Replicable functional magnetic resonance imaging evidence of correlated brain signals between physically and sensory isolated subjects. *J Altern Complement Med* 2005;**11**(6):955–63.
15. Hyman R, Honorton C. A joint communiqué: the psi ganzfeld controversy. *J Parapsychol* 1986;**50**:351–64.
16. May EC, Marwaha SB, Fallon JH. *Extrasensory perception: support, skepticism, and science.* Praeger; 2015.
17. Stanford RG, Palmer J. Free-response ESP performance and occipital alpha rhythms. *J Am Soc Psych Res* 1975;**69**(3):235–43.
18. Moulton ST, Kosslyn SM. Using neuroimaging to resolve the Psi debate. *J Cogn Neurosci* 2008;**20**(1):182–92.
19. Venkatasubramanian G, Jayakumar PN, Nagendra HR, Nagaraja D, Deeptha R, Gangadhar BN. Investigating paranormal phenomena: functional brain imaging of telepathy. *Int J Yoga* 2008;**1**(2):66–71.
20. Achterberg J, Cooke K, Richards T, Standish LJ, Kozak L, Lake J. Evidence for correlations between distant intentionality and brain function in recipients: a functional magnetic resonance imaging analysis. *J Altern Complement Med* 2005;**11**(6):965–71.
21. Roll WG, Persinger MA, Webster DL, Tiller SG, Cook CM. Neurobehavioral and neurometabolic (SPECT) correlates of paranormal information: involvement of the right hemisphere and its sensitivity to weak complex magnetic fields. *Int J Neurosci* 2002;**112** (2):197–224.
22. Persinger MA, Roll WG, Tiller SG, Koren SA, Cook CM. Remote viewing with the artist Ingo Swann: neuropsychological profile, electroencephalographic correlates,

magnetic resonance imaging (MRI), and possible mechanisms. *Percept Motor Skills* 2002;**94**(3 Pt 1):927–49.
23. Badgaiyan RD, Posner MI. Priming reduces input activity in right posterior cortex during stem completion. *Neuroreport* 1996;**7**(18):2975–8.
24. Badgaiyan RD, Posner MI. Time course of cortical activations in implicit and explicit recall. *J Neurosci* 1997;**17**(12):4904–13.
25. Badgaiyan RD, Schacter DL, Alpert NM. Auditory priming within and across modalities: evidence from positron emission tomography. *J Cogn Neurosci* 1999;**11**(4):337–48.
26. Schacter DL, Badgaiyan RD, Alpert NM. Visual word stem completion priming within and across modalities: a PET study. *Neuroreport* 1999;**10**(10):2061–5.
27. Badgaiyan RD. Neuroanatomical organization of perceptual memory: an fMRI study of picture priming. *Hum Brain Mapp* 2000;**10**(4):197–203.
28. Badgaiyan RD, Schacter DL, Alpert NM. Priming within and across modalities: exploring the nature of rCBF increases and decreases. *Neuroimage* 2001;**13**(2):272–82.
29. Schacter DL, Badgaiyan RD. Neuroimaging of priming: new perspectives on implicit and explicit memory. *Curr Direct Psychol Sci* 2001;**10**:1–4.
30. Badgaiyan RD, Schacter DL, Alpert NM. Priming of new associations: a PET study. *Neuroreport* 2003;**14**(18):2475–9.
31. Badgaiyan RD. Conscious awareness of retrieval: an exploration of the cortical connectivity. *Int J Psychophysiol* 2005;**55**(2):257–62.
32. Badgaiyan RD. Conscious awareness and the brain processing. *Elements* 2005;**3**:8–12.
33. Badgaiyan RD. Cortical activation elicited by unrecognized stimuli. *Behav Brain Funct* 2006;**2**(17):1–5.
34. Tigunait R, Patañjali. *The secret of the Yogasutra: Samadhi Pada*. Himalayan Institute; 2014.
35. Patañjali, Dvivedi MN, Subrahmanya Sastri S. *The Yoga-sūtras of Patañjali*. Theosophical Publishing House; 1934.
36. Bell IR, Schwartz GE. Voluntary control and reactivity of human heart rate. *Psychophysiology* 1975;**12**(3):339–48.
37. Curran EA, Stokes MJ. Learning to control brain activity: a review of the production and control of EEG components for driving brain-computer interface (BCI) systems. *Brain Cogn* 2003;**51**(3):326–36.
38. Maslach C, Marshall G, Zimbardo PG. Hypnotic control of peripheral skin temperature: a case report. *Psychophysiology* 1972;**9**(6):600–5.

Chapter 11

Dreams

Dreams have always fascinated and amazed people since time immorial. Their origin and significance are as mystifying today as they were in ancient times. There is a long history of our fascination with dreams. Right in the beginning of human civilization concepts about their significance began appearing in ancient literatures. Some philosophical concepts developed in the ancient world are surprisingly similar to what modern science is discovering now.

Ancient Indian scriptures are probably the oldest records of dream interpretation. Dreams, or *swapna,* are mentioned in *Rig Veda*, which dates to around 1500 BCE but existed in oral form much earlier.[1] While describing different kinds of dreams *Rig Veda* differentiates normal dreams from nightmares and dreaming while awake, which incidentally was considered evil. The concepts of *Rig Veda* were further developed in *Atharva Veda,* which dates back to 1200 BCE. In this scripture dreams were further characterized depending on the time of occurrence and nature of the dreamer. This paved the way for formulation of a comprehensive concept described in the *Upanishadas around 700 BCE*. In this collection of ancient scipture dreams are linked to consciousness. Even though different *Upanishadas* have slightly different concepts, all of them agree that dreaming is a part of our consciousness. *Chandokya Upanishada* describes four states of personal consciousness: *Jagrat* (waking); *swapna* (dream); *supta* (deep sleep); and *Turya* (fourth stage). This concept was further developed in the *Manduk Upanishada* in which the *Turya* is considered a form of self-consciousness that stays in one of three states: awake, dream, and deep sleep. It describes dreaming as a state in which the soul is immersed in its own light called *taijas*. According to this theory dreams are expressions of our inner desires. Yogic concepts developed later explain how dreams are controlled by self-consciousness. Interestingly, these concepts do not discount the possibility that the whole universe is a dream—the dream of the god *Brahma*. In parallel, another ancient Indian text the *Ayurveda* (scripted about 500 years BCE), which is a book of health and medicine, linked dreams with health. The two legends of this ancient medical system, Sushruta and Dhanvantari (Fig. 11.1), thought that dreams foretell our health. They described how different kinds of dreams indicate different aspects of health including recovery from an illness, deterioration of health, impending illness,

Neuroscience of the Nonconscious Mind. DOI: https://doi.org/10.1016/B978-0-12-816115-9.00011-5

FIGURE 11.1 Sushruta (top) and Dhanvantari (bottom). Legends of ancient medical system *Ayurveda*. Both of them associated dreams with health. *Images from Wikimedia Commons.*

sex of unborn child, etc. The *Ayurveda* even suggests methods to avoid bad outcome predicted by dreams.

Around the time ancient Indians started studying dreams, another civilization was independently developing its own theory.[2] People in Masopotamia located between the Tigris and Euphrates rivers studied dreams thoroughly and systematically documented meanings of different kinds of dreams. Masopotamians believed that dreams are sent by the god to tell us our future in symbolic form. They had "dream priests" to interpret dreams and control them using rituals. Masopotamian interpretation was popular throughout the Middle East in ancient times. Even now some cultures still believe in those interpretations.

Ancient Egyptians also had their book of dreams, which is now archived in the British Museum. Their theories were heavily influenced by Masopotamian concepts, which is why they also had dream interpreters who used to live in temples. They used to worship the dream god and sacrifice animals so that they can get dreams that accurately predict the future.

The Masopotamian concepts were also popular in Greece. There, two of the great thinkers, Aristotle and Hippocrates (Fig. 11.2), modified those ideas. Aristotle considered dreams a source of pure thought, and Hippocrates thought that they indicated physical and mental health. Both of them moved the origin

FIGURE 11.2 Greek thinkers: Top: Aristotle (384–322 BCE); Bottom: Hippocrates (460–370 BCE). Both of them moved origin of dreams from the God to an individual. *Images from Wikimedia Commons.*

of dreams away from gods and toward individuals. Their concepts are included in an influential ancient book called *Oneirocritica* (Interpretation of Dreams), written by Artemidorus.[3] Even though Aristotle and Hippocrates did not think that dreams tell the future, the Romans firmly believed in this ability. They had so much faith in the ability to predict the future through dreams that Emperor Augustus passed a law requiring citizens to report their dreams.

Interestingly, both Indian and middle eastern civilizations began studying dreams around the same time and both civilizations associated them with health also around the same time period. It may not be a coincidence that similar thoughts originated at the same time in the two civilizations, which were not known to have extensive philosophical exchanges in ancient times. it may suggest sequential nature of human intellectual evolution.

Even though philosophers and scientists have studied dreams for a long time, it is still a poorly understood phenomenon. After a flood of concepts developed in ancient times, no new concepts were developed for centuries. Then, toward the end of the nineteenth century, ancient Indian concept which considers dream as a part of personal consciousness, was restated with a few modification by Sigmund Fraud (Fig. 1.6) who published an influential book called *The Interpretation of Dreams* (German: *Die Traumdeutung*) in 1899.[4] At the time of publication his book did not get much attention, but today it is regarded as one of Freud's most significant works because he introduced the concept of unconscious mind in this book.

Apparently, Freud's idea on dreams originated in his dream, famously called "Irma's injection". Irma was one of Freud's patients who was not doing well. On the night of July 23, 1895 Freud had a dream in which he met Irma at a party. She looked weak and frail. Freud told her that she herself was responsible for deteriorating health because she did not follow his advice. In the dream Freud saw that Irma had a lesion in the throat and one of Freud's friends gave her an injection of a dangerous chemical using a "dirty" syringe. Freud woke up and thought this dream was meaningful. He immediately wrote it down and interpreted it as fulfillment of his wish that "he is not responsible for Irma's deteriorating health." Freud thought that he had uncovered the mystery of dream. In a letter to a friend he wrote, "One day there will be a marble tablet in this house saying: In this house on July 24, 1895, the secret of dreams was revealed to Dr. Sigm. Freud." It actually did happen. A tablet was put up with that inscription in front of the house in Austria where Freud had the dream (Fig. 11.3).

FIGURE 11.3 Memorial plate in commemoration of the place where Freud had the dream about "secrets of dreams" near Grinzing, Austria. *Modified from Wikimedia Commons.*

Freud suggested that dreams represent our unfulfilled desires that come from unconscious mind. These desires are modified by moral values and societal norms. According to Freud dreams have two types of content: manifest and latent. Manifest content is what we see and latent content is the underlying meaning we do not see or easily understand. In Freud's opinion every dream has a connection to events of the previous day, but the dream originates from one of the four sources: significant mental event, several mental events combined into one dream, past mental event triggered by a recent experience, and internal thought or memory. Freud interpreted many of his own dreams and explained their latent contents in the book. He used to interpret dreams of his patients to understand their unconscious conflicts.

Dreams have fascinated artists, poets, and writers throughout the history. There are thousands of classic paintings, poems, and novels on dream. Some of them are popular and thought provoking. Many artists including Salvador Dali (Fig. 8.3) and Picasso have painted their own dreams (Fig. 11.4). Scriptures on dreams are found in temples all over the world. They are particularly common in South Asia. Most famously, dream of the birth of Buddha was a popular theme of painting and sculpture. Apparently, Buddha's mother Maya in her dream saw a white elephant entering her body. When a saint was asked to interpret, he said Maya would soon give birth to a baby boy who would conquer the world. Ten months later she gave birth to a boy who would become Buddha. A sculpture depicting her dream was carved in many Buddhist complexes, including the one constructed in the third century BCE in Sanchi near Bhopal in India (Fig. 11.5).

There are many popular stories and novels based on dreams. My favorite is a story attributed to a Chinese philosopher Zhuangzi (369-286 BCE). In the story "Dream of a Butterfly" he articulates reality and illusion using dream as an example. The story goes like this: Once upon a time I, Zhuangzi, dreamt I was a butterfly, fluttering hither and thither to all intents and purposes a butterfly. I was conscious only of my happiness as a butterfly, unaware that I was Zhuangzi. Soon I awakened, and there I was, veritably myself again. Now I do not know whether I was then a man dreaming I was a butterfly, or whether I am now a butterfly, dreaming I am a man.[5] The story beautifully illustrates how dreams use all of our sensory faculties and make us feel that events in the dream are real. Of course, the story has a much deeper philosophical message concerning our desires and illusions.

In addition to being a subject of philosophical discussion, dreams have helped many of us solve problems in real life. Additionally, there is a long list of scientific breakthroughs that originated in dreams. The list includes discovery of the structure of benzene molecule in the dream of Friedrich August Kekulé, a German chemist. He reportedly dreamed about atoms dancing and a snake eating its own tail. This led him to understand the structure of a benzene molecule (Fig. 11.6). The double-helix structure of DNA also supposedly came in a dream of James Watson who saw DNA in the

FIGURE 11.4 Top: Salvador Dali's painting (1944) "Dream Caused by the Flight of a Bee Around a Pomegranate a Second Before Awakening" Bottom: Picasso's painting (1932). *"The Dream" Images from Wikimedia commons.*

FIGURE 11.5 Dream of Buddha's mother sculptured in Sanchi stupa, India. *Image from Wikimedia Commons.*

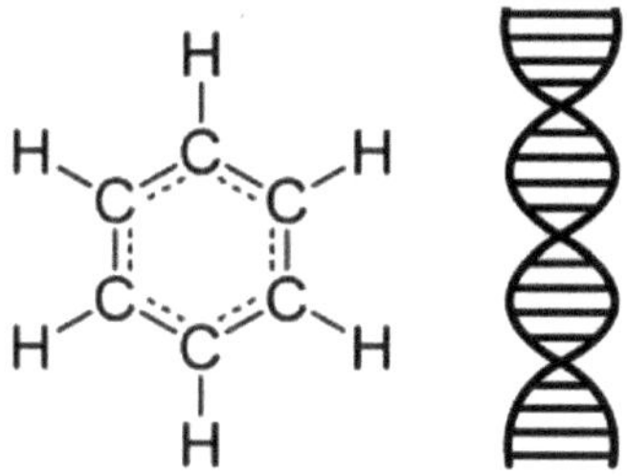

FIGURE 11.6 Structure of benzene molecule (left) and double-helix structure of DNA (right) were inspired by dreams.

form of a spiral staircase (Fig. 11.6). Similarly, Descartes got the idea of a "new science" in his dream, and the plot of the popular novel *The Strange Case of Dr Jekyll and Mr. Hyde* originated in a dream of the novelist Robert Louis Stevenson.

Dreams are particularly suited for generating creative ideas because they do not follow usual rules of associating objects and concepts. It is not uncommon to dream of unusual associations either between people or between people and places. Because of these unusual associations novel perspectives emerge. Dreaming allows us to "see" what we cannot visualize otherwise. Associations that we make while awake limit the range of our thought process. Dreams help us expand our thinking by loosening those associations.[6] As discussed in Chapter 8, Creativity, the ability to make unusual associations is an important quality of creative people. By making unusual association dreams help us find creative solution to our problems and anxieties.[7]

But we do not have novel insights in all dreams and most dreams do not express our creative genius. So why do we dream? Where do dreams come from? and How do they make our lives better, if they do?

To answer these questions many laboratories are studying dreams using behavioral, cognitive, and neuroimaging techniques. It is now a separate branch of science called oneirology. In the following paragraphs, I have discussed some of these experiments. This discussion is not comprehensive and it is narrowly focused on the nonconscious aspect of dreaming.

Early scientific study on dreams was conducted using electroencephalogram (EEG), which records electrical activity of the brain. In these studies volunteers were awakened repeatedly during sleep. It was found that most of our dreams come during a specific phase of sleep called rapid eye movement, or the REM phase. In this phase eyes move rapidly and high-frequency, low-amplitude beta waves dominate the EEG. These waves are normally seen while a person is awake with eyes open (Fig. 11.7). Dreams occur less frequently during the nonrapid eye movement, or NREM phase. In this phase the eyes do not move and EEG is dominated by low frequency

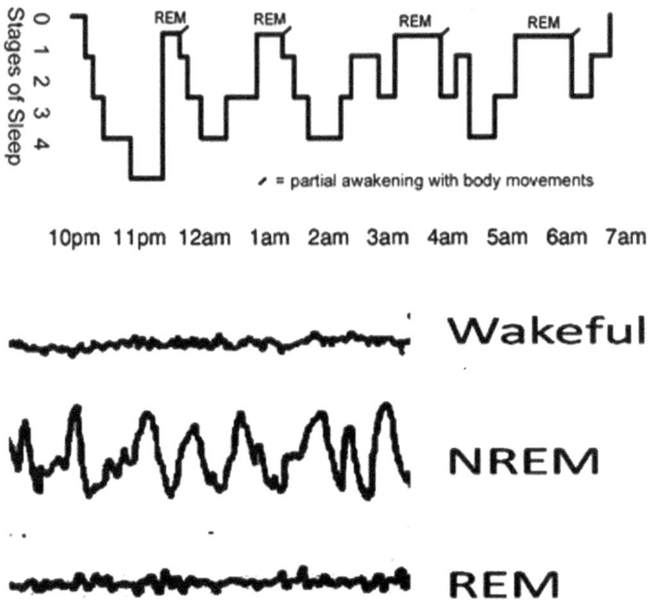

FIGURE 11.7 Top: Stages of sleep and approximate duration. Bottom: EEG recording at different stages: wakeful, NREM phase, and REM phase of sleep.

and high amplitude (theta and delta) waves. Most dreams that occur during NREM sleep are less vivid,[8] but vivid and terrifying night terrors also occur in this phase. Night terrors are common in children who wake up in terror usually a few minutes after falling sleep. NREM sleep comprises approximately 75% of our sleep time. This phase is further divided in four stages based on the depth of sleep. Stage 4 is the deepest and during this stage EEG waves are the slowest (Fig. 11.7).

We spend about 25% of sleep time in the REM phase and have 3–5 dreams every night, each lasting between a few seconds to 30 minutes.[8] Since most dreams occur during REM sleep we are more likely to remember the dream if we wake up during this phase.

Observation of an association between REM sleep and dreaming suggest that humans are not the only dreamers because all mammals and many non-mammals have REM sleep. So, do animals dream? Surprisingly, the answer is yes. Studies suggest that many animals including monkeys, elephants, cats, dogs, rats, and birds do actually dream.[9] But we do not know what they dream about.

If dream is so common across animals, it must be important biologically. Investigators are trying to understand what biological function or functions do dreams serve. The best approach to understand its function is to deprive animals and volunteers of dream and observe the effect. While it is difficult to deprive anyone of dreams it is possible to deprive people of REM sleep. By depriving REM sleep volunteers are also deprived of dreams because most (but not all) dreams occur in this phase. In REM deprivation studies volunteers are asked to sleep in the laboratory with EEG electrodes placed on the head. As soon as the EEG changes to the REM pattern, they are awakened either manually or by computer-generated signals. These studies have found that after REM deprivation people compensate for the loss by spending more sleep time in the REM phase, reducing the NREM time when they get a chance to sleep. This phenomenon is called REM rebound, and it makes volunteers irritable. They also have difficulty focusing attention and at times begin hallucinating.[10] But REM deprivation does have a positive effect. It lifts mood and mitigates depressive symptoms. The mood-elevating effect however, does not last long. After REM loss is compensated by REM rebound, the effect dissipates.[11] It is not a coincidence that many drugs that are used to elevate mood—either clinically or recreationally—alter REM sleep. These drugs include antidepressants such as fluoxetine (*Prozac*), and recreational drugs like cocaine and amphetamine. It is not clear if altered REM sleep contributes to the mood-elevating effect of these substances, but this possibility cannot be discounted.

Because of the adverse consequences, a volunteer can be REM deprived only for a limited period of time. Therefore, studies on the effects of long-term REM deprivation are conducted on laboratory animals. These studies have consistently reported early deaths. A Chicago sleep medicine expert named Clete Anthony Kushida reported that his REM sleep deprived rats died in 16–54 days despite increased food and adequate water intake.[12] He subsequently found that these deaths were due to enhanced energy expenditure. These animals consume almost twice as much energy as normal rats. It is an interesting finding but leaves an unanswered question – why do they have higher energy consumption?

It is not known how REM sleep affects energy expenditure but we do know that sleep is important for consolidation of memory.[13] Further, different types of memory are consolidated during different sleep phases. Most conscious or explicit memories are consolidated in the NREM phase and consolidation of nonconscious or implicit and procedural memories (e.g., cycling, typing) occurs during REM sleep.[14] The role of sleep or dreams in memory consolidation is unclear. We do not know whether dreams somehow help the consolidation process or they are the product of consolidation process.

During memory consolidation new information is linked to old memories. In the process of "finding" the right piece of memory to link to newly acquired information, many brain areas are activated. It is possible that those activations are perceived as dreams. But if that is the case, then why do we dream more often in the REM phase? Memories are consolidated in the NREM phase too. One possible explanation is the lack of "action" in explicit memories that are consolidated in the NREM phase. Lack of action prevents the brain from making "storylines" for dreams. This argument is supported by the fact that people experience 'mentation' which refers to some form of thought process during NREM sleep.[15–17] Mentation is reported in 43% of awakenings during this phase. Obviously, these thoughts are not dreams but their existence suggests that the process of memory consolidation could be activating areas of the brain causing either dreams or mentation depending on the type or content of the memory being consolidated. Some explicit memories do have motor components and those memories do create dreams in the NREM phase. This is why 5%–10% of awakened during this phase report vivid dreams similar to those reported during REM sleep.[18]

This discussion does not imply that REM sleep is necessary for memory consolidation. It is not. People who have damage to the brainstem do not have REM sleep but consolidation of their memory remains unaffected.[19] Loss of REM sleep following damage to the brainstem indicates that REM sleep might originate in this part of the brain. This connection was first made in the mid-1970s by two Harvard psychiatrists, Allan J. Hobson and Robert W. McCarley (Fig. 11.8). They found that REM and NREM sleep are controlled by separate group of brainstem neurons they called REM-on and REM-off neurons. Activation of REM-on neurons initiates REM sleep, which is terminated when REM-off neurons fire. Termination of REM sleep coincides with the beginning of NREM sleep. They also found that activity of these neurons is controlled by different neurotransmitters. REM-on neurons are controlled by the cholinergic system and REM-off by serotonergic and noradrenergic systems.[20–22]

These observations led Hobson and McCarley to formulate the activation-synthesis theory of dreams.[21,22] It suggests that dreams originate in the brainstem and activate higher brain areas involved in processing of emotion and memory, particularly the limbic cortex, hippocampus, and amygdala. According to the theory, activation in these areas is synthesized

FIGURE 11.8 Allan Hobson (born 1933) and Robert McCarley (1937–2017) proposed the activation-synthesis theory of dream. *Pictures courtesy of Dr Hobson and Dr McCarley's family.*

and this synthesis is interpreted by the brain as a dream. It provided a good starting point for understanding the neural networks that control dreams, but this theory was based on the assumption that dreams occur only in the REM phase. It implied that the brain activity associated with REM sleep represents dream-related activity. We now know that this assumption is not true. Most but not all dreams occur in the REM phase. Some of them also occur in NREM sleep. But areas of the brain Hobson and McCarley associated with REM sleep/dream are accurate because recent neuroimaging experiments have reported activity in some of the same areas (Fig. 11.9).[23] These activations, we now know, are associated with REM sleep not with dream.

Doubts about validity of the activation-synthesis hypothesis were raised when it was observed that people with damaged brainstems lose REM sleep but not dreams.[16,24] It suggested that REM sleep and dream are processed by different neural mechanisms. This idea got further support with the observation that lesions in certain cortical areas cause selective loss of dream while preserving REM sleep.[15,16,25,26] Further, it was observed that 5%–30% of people awakened during REM phase do not report dreaming and 5%–10% of those in the NREM phrase report dreams that are indistinguishable from REM dreams.[18] These observations made it clear that dream and REM sleep are controlled by separate neural networks.

The activation-synthesis theory was therefore modified primarily after the studies of a South African psychoanalyst and dream researcher Mark Solms. He studied dreams of hundreds of brain-damaged patients in neurosurgery department of University of Cape Town and found that people with lesions in the parietal cortex do not dream but those with lesions in the brainstem do.[16] Based on this observation he concluded that the brainstem is not the site of origin of dreams and that the higher brain areas are involved in its processing. This observation prompted the study of brain areas that might be

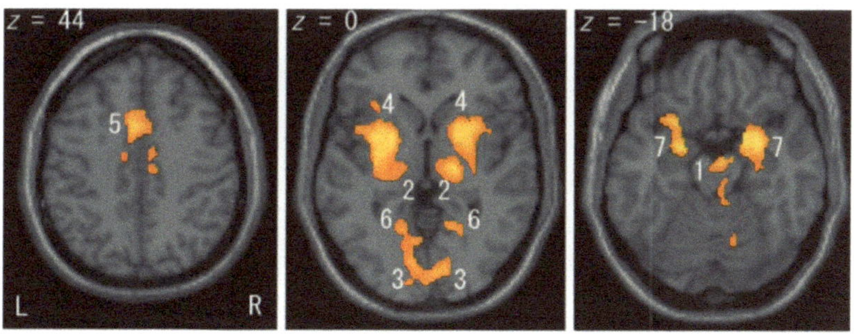

FIGURE 11.9 fMRI images of the brain areas activated during REM sleep: 1 tegmentum, 2 thalamus, 3 visual cortex, 4 putamen, 5 anterior cingulate, 6 parahippocampus, and 7 amygdala. *Modified from Miyauchi S, Misaki M, Kan S, Fukunaga T, Koike T. Human brain activity time-locked to rapid eye movements during REM sleep.* Exp Brain Res *2009;**192**(4):657–67. Picture courtesy Dr Miyauchi.*

associated with dream. One study found that vividness of dreams correlate with the volume of hippocampus, amygdala, and medial prefrontal cortex[27] and stimulation of the amygdala and hippocampus elicits a dream-like response.[28] These and other findings suggest that multiple brain areas are involved either in initiation or enhancement of dreams. These areas include the orbitofrontal cortex, occipital cortex, dorsolateral preorbital cortex, anterior cingulate cortex, medial prefrontal cortex, and area at the junction of the occipital, parietal, and temporal lobes (OPT junction).[16] Out of these, the last two areas are most important (Fig. 11.10) because after lesions in these areas people do not recall dreams.[16] Other areas modify dreams by making them vivid and colorful. Activation of the dorsomedial prefrontal cortex enhances vividness of dreams and the amygdala and limbic structures are activated when dreams are emotionally charged. Interestingly, activity in the primary visual cortex is reduced during dream and lesions in this area do not affect dream or its recall, even though it makes a person blind.[15] On the contrary, people who have lesions in the visual association areas cannot recognize faces or see colors both in real life and in dreams.[15]

Involvement of the medial prefrontal cortex and the cortical area around OPT junction in dream initiation and recall was confirmed both by lesion and neuroimaging studies. It was found that people are unable to recall dreams after lesions around the OPT junction[16,25] or in the medial prefrontal cortex.[29] Similarly, complete loss of dream is reported after stroke of the posterior cerebral artery that supplies the OPT junction.[26] Additionally, people reported losing dream after a modified prefrontal leucotomy that involves sectioning the white matter fibers that connect the medial prefrontal cortex to rest of the brain.[30] This procedure was performed on thousands of psychiatric patients in the 1940s and 1950s to make them less aggressive. Even though the Portuguese neurologist, Egas Moniz who first suggested the procedure, was awarded the Nobel Prize, the procedure was abandoned because of the emotional and intellectual impairments associated with it.

Involvement of the OPT junction and medial prefrontal cortex in dream was confirmed by neuroimaging studies. In a study conducted by J.B.

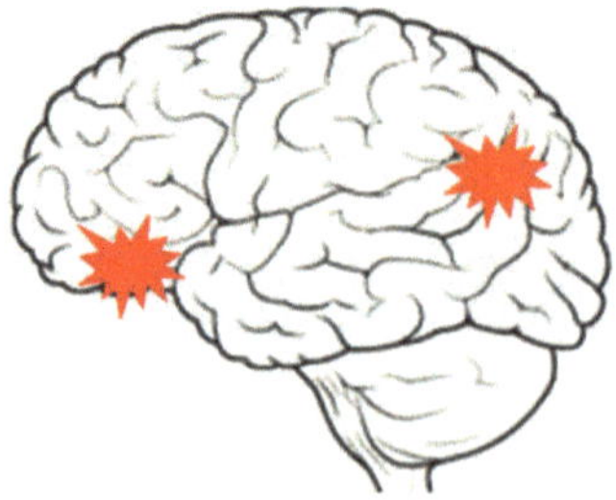

FIGURE 11.10 Damage to the marked areas of the brain (medial prefrontal cortex and OPT junction) impair recall of dreams.

Eichenlaub and colleagues in Lyon France, positron emission tomography (PET) was used to study changes in the regional blood flow of the brain during dreaming in two groups of volunteer.[31] One group comprised of people who were good at recalling dreams (high recallers) and the other of volunteers who were not so good (low recallers). These volunteers were first sleep deprived and then asked to sleep inside a scanner. Changes in the regional cerebral blood flow were measured while they slept. As expected, about 90% of the high and 40% of low recallers were able to recall their dreams while sleeping inside the scanner. Analysis of the data revealed that high recallers had higher blood flow in the medial prefrontal cortex and OPT junction (Fig. 11.11).

We know that the brain mechanisms of memory retrieval are the same in both awake and dream states[32,33] and the prefrontal cortex is involved in the retrieval of both memory and dreams.[32,33] Therefore, the inability to recall dreams following a lesion in the prefrontal cortex could be due to impaired retrieval rather than the absence of dreaming.

It then leaves the occipito-parieto-temporal (OPT) junction as the site where dreams are initiated. Many researchers have confirmed that a lesion in this area completely abolishes all dreaming activity.[16,25] This observation makes an interesting connection between dreams and our previous experiments. In Chapter 2, Nonconscious memory, we discussed a series of neuroimaging experiments conducted to understand the neural basis of nonconscious memory. We found that area V3A, which is located in the

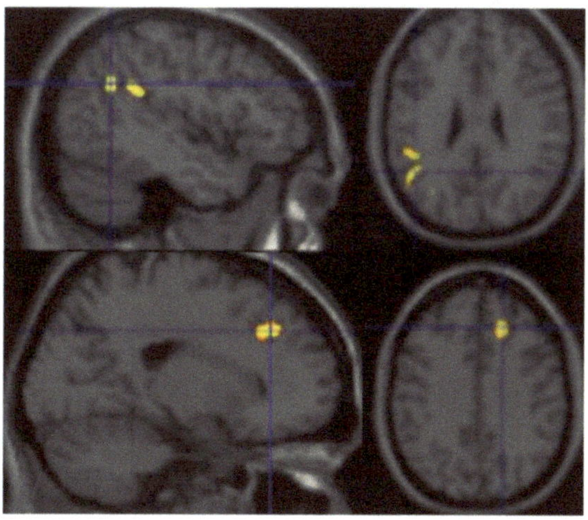

FIGURE 11.11 Increased activation (*marked yellow*) in the occipito-parieto-temporal (OPT) junction (top) and in the medial prefrontal cortex (bottom) in high dream recallers as compared to low recallers. *Adopted from 1 Eichenlaub JB, Nicolas A, Daltrozzo J, Redoute J, Costes N, and Ruby P. (2014). Resting brain activity varies with dream recall frequency between subjects.* Neuropsychopharmacology *39(7), 1594–1602.*

OPT junction, is critically involved in processing of nonconscious memory.[34–37] We observed reduced activation in this area during retrieval without conscious awareness. [34–36,38–41] This observation is significant in view of the dream-induced activation reported in the PET study mentioned above.[31] In this study, Eichenlaub and colleagues found reduced activity in the same OPT area during dreaming where we found reduction during nonconscious retrieval. Since reduced activity indicates nonconscious retrieval, dream probably retrieve information from nonconscious pool. Further, as explained in Chapter 2, Nonconscious memory, area V3A has a gating mechanism that allows retrieval of a limited amount of relevant information.[34–37] Further, the degree of reduction in V3A activity is inversely proportional to the amount of information released. Therefore, greater reduction suggests greater restriction resulting in release of relatively small amount of information. It thus makes perfect sense that low dream recallers had greater reduction in Eichenlaub's study.[31] Essentially, greater reduction makes them low recallers. On the other hand, reduction in high recallers was relatively small (therefore, a relative increase in activation as compared to the activation in low recallers), suggesting release of a larger pool of information in their dreams. It made their dream content rich and allowed better recall.

Interestingly, this observation is consistent with ancient theories of dream. As discussed earlier, in ancient Indian literature dream is considered a stages of consciousness.[1,42] It does appear to be a part of nonconscious-conscious continuum. As discussed above, dreams share features of conscious experience and also those of nonconscious processing. For example, dreams use all of our sensations to see, hear, smell, touch, and taste objects just like we do while awake and fully conscious. Dream also maintains our personality with its strengths and weaknesses and reflects the mood, imaginativeness, interests, and concerns of our waking life.[43] While maintaining these characteristics of conscious state, dreams are nonconscious processes not only because we are not awake while dreaming, but also because of reduced activity in the OPT/V3A area. As discussed above, this reduction is a characteristic feature of nonconscious processing.[34,35,40,41,44]

As discussed in Chapter 12, Nonconscious mind: smart or dumb, both conscious and nonconscious processing bring strength to our mental abilities. So, how do dreams enhance our abilities? We already discussed how dreams help us find creative solutions to our problems by making unusual associations. This, however, does not happen very often even though we dream every day. Dreams therefore must have other functions.

One important function of dreams is to "make stories." Dreams make stories around our problems, sometimes explicitly but mostly in a subtle way. Once a problem that bothers us becomes a story, our anxiety dissipates to a great extent because stories by definition are things of the past. Problems in real life cause anxiety because of their future consequences. By making them a story, dreams tell us subtly that we have left the problem behind. This thought is further reinforced by memory. Memories of dreams is processed and retained in the brain in the same way as memories of real events. By retaining the dream as a "story of the problem" in memory, the brain again gives an impression that the problem is a thing of the past because memories are recollection of past events. Thus, dreams help us redefine problems and help reduce anxiety about their future consequences. In addition, the brain can handle bad memories better than the anxiety associated with future unknown consequences. Additionally, dreams help us trivialize problems by repeated presentation and making them a part of a story. When we "see" our problem in dream as a story over and over again, the problem becomes "familiar," and as we know, familiarity brings comfort and reduces anxiety[45] as discussed in Chapter 7, Emotion.

Thus, an important function of dream is to reduce anxiety and help us face problems effectively. This function was demonstrated in an interesting experiment conducted by a Chicago psychologist Rosalinda Cartwright. She studied the dreams of 70 women who were going through divorce. They were asked to sleep in the laboratory for three nights and record their dreams. It was observed that depressed women generally had negative dreams involving ex-spouses more often. After a year of follow up Cartwright found that women who had dreams of their spouses were better adjusted and were in a much better mental state than those who did not.[46,47] This finding demonstrates how dreams help us cope up with our problems and anxieties.

Similarly, after experiencing life-threatening trauma many individuals experience nightmares on a regular basis. While unpleasant, these dreams help reduce the mental impact of trauma. We know that repeated exposure makes trauma less threatening. This technique, called exposure therapy, is used to treat phobias. In this therapy the object of fear is exposed to a phobic individual gradually over several days until the individual has no fear of the object. Similarly, people with a history of trauma are exposed to the traumatic situation in progressively increasing severity until the trauma ceases to elicit anxiety. Nightmares serve the same purpose. By repeated exposure of trauma in dreams, fear of experienced trauma either goes away or its intensity diminishes. This was experimentally demonstrated in a study on Palestinian children by a Finnish psychologist Raija-Leena Punamäki. She found that traumatized Palestinian children recall dreams more often and their dreams have more negative content. She also found that these dreams made them more resilient by "normalizing" the trauma.[48,49]

The fact that the predominant emotions[50] in dreams are joy, surprise, anger, fear, and anxiety also suggests that dreams help us cope up with our problems. These emotions allow us to look at different dimensions of the problem and provide optimism. Negative emotions that drag us down are rare in dreams, which is why we usually do not experience emotions such as sadness, guilt, and depression in our dreams.

Dreams also allow us to look at problems rationally because about 25% of dreams do not have any emotional component and do not generate emotion in situations that would be emotional while awake.[51] We know that looking at a problem without emotion sometimes helps. By removing emotion, dreams can help find solutions to problems. Since separating emotion from problems is virtually impossible while awake, dreams do it for us while we sleep. Dreams therefore work as natural psychotherapy. They help us solve problems and recover from trauma. It may be the last line of defense against adverse circumstances. Dreams make us better prepared to face difficult situations and make us resilient.

Like other nonconscious functions there is individual variation in our ability to use dreams to help solve problems. Some of us experience an anomalous reaction. In those individuals, instead of helping, dreams hurt them mentally. Individuals with posttraumatic stress disorder (PTSD) are examples of this form of reaction. In these individuals, nightmares instead of "trivializing" memories of trauma retraumatize over and over again. In these people, repeated nightmares increase anxiety. But this happens in only a small fraction of individuals who develop PTSD after a traumatic experience. Nightmares therefore affect people either positively or negatively. Those experiencing positive effects recover from trauma while people who experience negative effects experience PTSD symptoms. We do not know exactly what makes individuals respond differently to nightmares but one of our experiments provides a clue by delineating the difference in the way emotions are processed in healthy people and in individuals with PTSD symptoms.

People with PTSD have a different response to nightmares possibly because they have difficulty processing emotions triggered by traumatic events. In order for dreams to provide relief by repeated exposure, it is important that emotions be fully processed. Only a processed emotion can make the brain understand its significance and initiate appropriate dreams and other coping mechanisms. Partially processed emotions cannot respond to these natural damage control strategies. Since emotions are processed by the dopaminergic system we studied dopamine neurotransmission during emotional processing in individuals with PTSD to understand if they have difficulty processing emotions.

This study is particularly significant because dopamine is involved in emotional processing as well as in initiation of dreams.[15] It also makes dreams lucid.[52] That is the reason most psychotic patients who have excessive dopamine in their system complain of vivid dreams[53] and medications that enhance dopamine such as levodopa and bupropion also have a similar effect.[54] More significantly, lesions in the dopaminergic tracks around the medial prefrontal cortex abolish dream recall.[16] Since the dopaminergic system is involved in processing of both emotion and dreams, dysregulation of the system would affect both.

To study dopamine neurotransmission we used the neurotransmitter imaging method, single scan dynamic molecular imaging technique (SDMIT) discussed earlier.[55–66] In this study, we focused on dopamine transmission in the basal ganglia because emotions are processed in this area before they are refined at the higher brain centers. The study included healthy control and PTSD volunteers. In the study, volunteers received intravenous bolus of a radiolabeled dopamine receptor ligand ^{11}C-raclopride after they were positioned in a PET camera. Then a series of emotionally neutral words (e.g., chair, table, park) were presented and volunteers were asked to press a key on a keypad based on the intensity of emotion elicited by each word (1 for no emotion and 3 for intense emotion). After 20 minutes, unbeknownst to them, stimuli were changed from neutral to emotional words (e.g., blood, gun, rape). Expectedly, these words generated more intense emotion. While volunteers were studying those words, PET data were acquired and the amount of dopamine released was measured dynamically based on concentration of the ligand in different brain areas. Since ligand binds to the same receptors as dopamine, it competes with endogenously released neurotransmitter for occupancy of receptor sites. As a result, endogenously released dopamine displaces ligand from receptor sites. The ligand concentration therefore reduces in areas where dopamine is released. Using this strategy we mapped and measured ligand concentration during presentation of emotionally neutral words and also when emotional words were presented. We found that dopamine displaced ligand from receptor sites reducing its concentration when emotional words were presented. In healthy volunteers the reduction was most significant in the caudate and putamen of both hemispheres (Fig. 7.5) suggesting release of dopamine in these areas during emotional processing. In individuals with PTSD the pattern of dopamine release was different. We found reduced concentration of the ligand only in the caudate and right putamen (Fig. 11.12), indicating reduced amount of dopamine release. We also found that the rate of dopamine release was slower in the caudate of PTSD volunteers. In healthy volunteers, it peaked within a minute of presenting emotional words but in PTSD it took two minutes to achieve the same peak (Fig. 11.12).

Because of reduced dopamine neurotransmission, emotional processing in PTSD is probably incomplete. This may be the reason dreams do not help people with PTSD cope with traumatic experience. Instead of helping them recover from trauma they continue to retraumatize. The best treatment strategy for these individuals therefore is to fix dysregulated dopamine neurotransmission. It would allow them to process emotions and their memories to completion. Their dreams would then help them recover instead of retraumatizing them repeatedly. Individuals with PTSD would then be able to use the natural healing power of dreams.

Dream therefore, can work as a healer and make us resilient. Its healing power could be further enhanced if we acquire the ability to control its

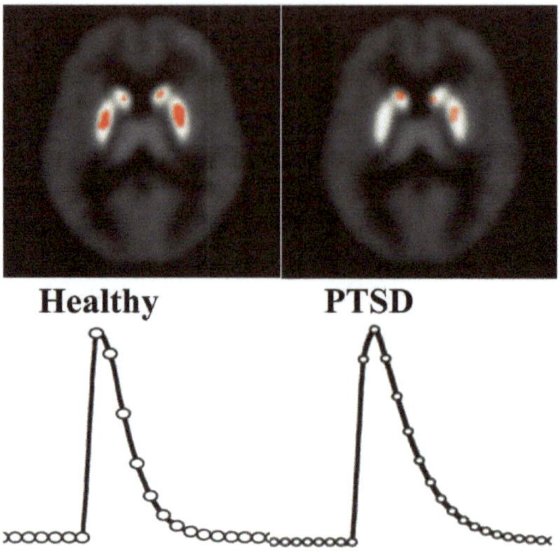

FIGURE 11.12 Areas of the caudate and putamen where dopamine was released (*red*) in healthy (left) and PTSD (right) volunteers. Graphs show slow rate of dopamine release during emotional processing in the caudate of PTSD volunteers.

contents. We are not there yet. At this time we have limited understanding of neural control of dreams. The observation that dreams are possibly regulated in area V3A (like other nonconscious processes) provides an important starting point to study the world of dreams. In-depth study of area V3A could in near future provide us the ability to control our dreams. If that happens, dreams could be used to enhance creativity, to learn coping strategies, and to treat psychiatric conditions. I am sure soon this dream will come true.

Bibliography

1. Tendulkar SS, Dwivedi RR. 'Swapna' in the Indian classics: mythology or science? *Ayu* 2010;**31**(2):170–4.
2. Hughes JD. Dream interpretation in ancient civilizations. *Dreaming* 2000;**10**(1):7–18.
3. Artemidorus, White RJ. *The interpretation of dream = Oneirocritica*. Noyes Press; 1975.
4. Freud S. *The interpretation of dreams*. Modern Library; 1899.
5. Wu K-M, Zhuangzi. *The butterfly as companion: meditations on the first three chapters of the Chuang Tzu*. State University of New York Press; 1990.
6. Fiss H, Klein GS, Bokert E. Waking fantasies following interruption of two types of sleep. *Arch Gen Psychiatry* 1966;**14**(5):543–51.
7. Wagner U, Gais S, Haider H, Verleger R, Born J. Sleep inspires insight. *Nature* 2004;**427** (6972):352–5.
8. Hobson JA. REM sleep and dreaming: towards a theory of protoconsciousness. *Nat Rev Neurosci* 2009;**10**(11):803–13.
9. Schredl M. Animal dreams in a long dream series. *Int J Dream Res* 2013;**6**(1):59–64.

10. Vogel GW. A review of REM sleep deprivation. *Arch Gen Psychiatry* 1975;**32**(6):749–61.
11. Vogel GW, Thurmond A, Gibbons P, Sloan K, Walker M. REM sleep reduction effects on depression syndromes. *Arch Gen Psychiatry* 1975;**32**(6):765–77.
12. Kushida CA, Bergmann BM, Rechtschaffen A. Sleep deprivation in the rat: IV. Paradoxical sleep deprivation. *Sleep* 1989;**12**(1):22–30.
13. Rasch B, Born J. About sleep's role in memory. *Physiol Rev* 2013;**93**(2):681–766.
14. Tucker MA, Hirota Y, Wamsley EJ, Lau H, Chaklader A, Fishbein W. A daytime nap containing solely non-REM sleep enhances declarative but not procedural memory. *Neurobiol Learn Mem* 2006;**86**(2):241–7.
15. Solms M. *The neuropsychology of dreams: a clinico-anatomical study*. L. Erlbaum Associates; 1997.
16. Solms M. Dreaming and REM sleep are controlled by different brain mechanisms. *Behav Brain Sci* 2000;**23**(6):843–50 discussion 904–1121.
17. McNamara P, Johnson P, McLaren D, Harris E, Beauharnais C, Auerbach S. REM and NREM sleep mentation. *Int Rev Neurobiol* 2010;**92**:69–86.
18. Hobson JA. *The dreaming brain*. Basic Books; 1988.
19. Siegel JM. The REM sleep-memory consolidation hypothesis. *Science* 2001;**294**(5544):1058–63.
20. Steriade M, Hobson J. Neuronal activity during the sleep-waking cycle. *Prog Neurobiol* 1976;**6**(3–4):155–376.
21. Hobson JA, McCarley RW. The brain as a dream state generator: an activation-synthesis hypothesis of the dream process. *Am J Psychiatry* 1977;**134**(12):1335–48.
22. McCarley RW, Hobson JA. The neurobiological origins of psychoanalytic dream theory. *Am J Psychiatry* 1977;**134**(11):1211–21.
23. Miyauchi S, Misaki M, Kan S, Fukunaga T, Koike T. Human brain activity time-locked to rapid eye movements during REM sleep. *Exp Brain Res* 2009;**192**(4):657–67.
24. Chase TN, Moretti L, Prensky AL. Clinical and electroencephalographic manifestations of vascular lesions of the pons. *Neurology* 1968;**18**(4):357–68.
25. Murri L, Massetani R, Siciliano G, Giovanditti L, Arena R. Dream recall after sleep interruption in brain-injured patients. *Sleep* 1985;**8**(4):356–62.
26. Bischof M, Bassetti CL. Total dream loss: a distinct neuropsychological dysfunction after bilateral PCA stroke. *Ann Neurol* 2004;**56**(4):583–6.
27. De Gennaro L, Lanteri O, Piras F, Scarpelli S, Assogna F, Ferrara M, et al. Dopaminergic system and dream recall: an MRI study in Parkinson's disease patients. *Hum Brain Mapp* 2016;**37**(3):1136–47.
28. Halgren E, Walter RD, Cherlow DG, Crandall PH. Mental phenomena evoked by electrical stimulation of the human hippocampal formation and amygdala. *Brain* 1978;**101**(1):83–117.
29. Yu C. Cessation of dreaming and ventromesial frontal region infarcts. *Neuropsychoanalysis* 2004;**9**(1):83–90.
30. Schindler R. [Dreaming in lobotomized persons]. *Wien Z Nervenheilkd Grenzgeb* 1953;**6**(4):330–4.
31. Eichenlaub JB, Bertrand O, Morlet D, Ruby P. Brain reactivity differentiates subjects with high and low dream recall frequencies during both sleep and wakefulness. *Cereb Cortex* 2014;**24**(5):1206–15.
32. De Gennaro L, Marzano C, Cipolli C, Ferrara M. How we remember the stuff that dreams are made of: neurobiological approaches to the brain mechanisms of dream recall. *Behav Brain Res* 2012;**226**(2):592–6.

33. Marzano C, Ferrara M, Mauro F, Moroni F, Gorgoni M, Tempesta D, et al. Recalling and forgetting dreams: theta and alpha oscillations during sleep predict subsequent dream recall. *J Neuroscience* 2011;**31**(18):6674–83.
34. Badgaiyan RD, Schacter DL, Alpert NM. Auditory priming within and across modalities: evidence from positron emission tomography. *J Cogn Neurosci* 1999;**11**(4):337–48.
35. Badgaiyan RD. Neuroanatomical organization of perceptual memory: an fMRI study of picture priming. *Hum Brain Mapp* 2000;**10**(4):197–203.
36. Badgaiyan RD, Schacter DL, Alpert NM. Priming within and across modalities: exploring the nature of rCBF increases and decreases. *Neuroimage* 2001;**13**(2):272–82.
37. Badgaiyan RD. Nonconscious processing and a novel target for schizophrenia research. *Open J Psychiatry* 2012;**2**(4A):335–9.
38. Badgaiyan RD. Conscious awareness of retrieval: an exploration of the cortical connectivity. *Int J Psychophysiol* 2005;**55**(2):257–62.
39. Schacter DL, Badgaiyan RD, Alpert NM. Visual word stem completion priming within and across modalities: a PET study. *Neuroreport* 1999;**10**(10):2061–5.
40. Badgaiyan RD, Posner MI. Time course of cortical activations in implicit and explicit recall. *J Neuroscience* 1997;**17**(12):4904–13.
41. Badgaiyan RD, Posner MI. Priming reduces input activity in right posterior cortex during stem completion. *Neuroreport* 1996;**7**(18):2975–8.
42. Rao KR. Anomalies of consciousness: Indian perspectives and research. *J Parapsychol* 1994;**58**(2):149–87.
43. Hall CS, Van de Castle RL. *The content analysis of dreams*. Appleton-Century-Crofts; 1966.
44. Schacter DL, Badgaiyan RD. Neuroimaging of priming: new perspectives on implicit and explicit memory. *Curr Direct Psychol Sci* 2001;**10**:1–4.
45. Zajonc RB, Reimer DJ, Hausser D. Imprinting and the development of object preference in chicks by mere repeated exposure. *J Compar Physiol Psychol* 1973;**83**(3):434–40.
46. Cartwright R, Young MA, Mercer P, Bears M. Role of REM sleep and dream variables in the prediction of remission from depression. *Psychiatry Res* 1998;**80**(3):249–55.
47. Cartwright R. Dreams and adaptation to divorce. In: Barrett D, editor. *Trauma and dreams*. Harvard University Press; 1996.
48. Valli K, Revonsuo A, Palkas O, Punamaki RL. The effect of trauma on dream content – a field study of Palestinian children. *Dreaming* 2006;**16**(2):63–87.
49. Punamaki RL. The role of dreams in protecting psychological well-being in traumatic conditions. *Int J Behav Dev* 1998;**22**:559–88.
50. Strauch I, Meier B. *In search of dreams: results of experimental dream research*. State University of New York Press; 1996.
51. Foulkes D, Sullivan B, Kerr N, Brown L. Appropriateness of dream feelings to dreamed situations. *Cogn Emotion* 1988;**2**(1):29–39.
52. Dey S, Hafkemeyer P, Pastan I, Gottesman MM. A single amino acid residue contributes to distinct mechanisms of inhibition of the human multidrug transporter by stereoisomers of the dopamine receptor antagonist flupentixol. *Biochemistry* 1999;**38**(20):6630–9.
53. Moskovitz C, Moses III H, Klawans HL. Levodopa-induced psychosis: a kindling phenomenon. *Am J Psychiatry* 1978;**135**(6):669–75.
54. Posner J, Bye A, Dean K, Peck AW, Whiteman PD. The disposition of bupropion and its metabolites in healthy male volunteers after single and multiple doses. *Eur J Clin Pharmacol* 1985;**29**(1):97–103.

55. Badgaiyan RD. Imaging dopamine neurotransmission in live human brain. *Prog Brain Res* 2014;**211**:165–82.
56. Badgaiyan RD. Detection of dopamine neurotransmission in "real time". *Front Neurosci* 2013;**7**:125.
57. Badgaiyan RD. Neurotransmitter imaging: current status and challenges. *Curr Med Imag Rev* 2011;**7**:96–8.
58. Badgaiyan RD. Neurotransmitter imaging: basic concepts and future perspectives. *Curr Med Imag Rev* 2011;**7**:98–103.
59. Badgaiyan RD. Dopamine is released in the striatum during human emotional processing. *Neuroreport* 2010;**21**:1172–6.
60. Badgaiyan RD, Fischman AJ, Alpert NM. Dopamine release during human emotional processing. *Neuroimage* 2009;**47**(4):2041–5.
61. Badgaiyan RD, Fischman AJ, Alpert NM. Explicit motor memory activates the striatal dopamine system. *Neuroreport* 2008;**19**(4):409–12.
62. Badgaiyan RD, Fischman AJ, Alpert NM. Striatal dopamine release in sequential learning. *Neuroimage* 2007;**38**(3):549–56.
63. Fischman AJ, Badgaiyan RD. Neurotransmitter imaging. In: Charron M, editor. *Pediatric PET*. Springer; 2006. p. 385–403.
64. Badgaiyan RD, Fischman AJ, Alpert NM. Detection of striatal dopamine released during an explicit motor memory task. *J Nucl Med* 2005;**46**(Suppl. 2):213.
65. Badgaiyan RD, Fischman AJ, Alpert NM. Striatal dopamine release during unrewarded motor task in human volunteers. *Neuroreport* 2003;**14**(11):1421–4.
66. Alpert NM, Badgaiyan RD, Livini E, Fischman AJ. A novel method for noninvasive detection of neuromodulatory changes in specific neurotransmitter systems. *Neuroimage* 2003;**19**(3):1049–60.

Chapter 12

Nonconscious mind: smart or dumb?

As discussed in previous chapters, the nonconscious mind differs from the conscious mind in a number of ways. Since we are not consciously aware of the reasoning behind our nonconscious decisions, it is generally believed to be a dumb system. Let us examine if nonconscious mind is really dumb.

The most important feature of the nonconscious mind is its speed of processing. It is quick and in many instances immediate.[1] Nobody doubts its superiority in situations where quick decisions and actions are needed. Most decisions that we make in real or perceived emergency situations are therefore nonconscious. Thus our immediate reaction to look toward and run away from a source of loud noise is a decision the nonconscious mind makes for us based on the knowledge that the loud noise is unusual and may not be safe. Similarly, if we see a person entering the room with an open knife, we nonconsciously react and immediately prepare ourselves for fight or flight. We do not spend time debating whether the person with the knife intends to hurt us or not. The nonconscious mind prepares for the worst and immediately removes us from a potentially dangerous situation. We tend to undermine its life-preserving function because those situations are not encountered every day and we do not even know that those decisions and actions are nonconscious in nature.

Quick nonconscious decisions therefore save lives. But it does not mean the nonconscious mind helps us only in rare life-threatening situations. We use it in our daily lives probably more often than the conscious mind. First, the nonconscious mind helps us perform activities that we have perfected by practice. The movement of fingers while typing and those of legs while walking are examples of activities we perform nonconsciously. We use the nonconscious mind not only while awake but also during sleep in the form of dreams. The most frequently used nonconscious function is probably linguistic processing. Almost all aspects of linguistic expression involve the nonconscious mind. Learning of a language is mostly nonconscious, particularly for the first language. After a language is learned, its use depends on nonconscious processing. Use of language involves processing of a large number of information in a relatively short period of time. Because of its

Neuroscience of the Nonconscious Mind. DOI: https://doi.org/10.1016/B978-0-12-816115-9.00012-7

faster processing speed, only the nonconscious mind can accomplish this task. We cannot have a conversation if language is not processed nonconsciously. Imagine how long would it take to consciously recall thousands of words we have learned and then choose the one most appropriate during a conversation. Moreover, in a sentence we use many words and also make sure the sentence is grammatically correct. Conscious recall of a five-word sentence and applying the rules of grammar would require several minutes at best. This latency would make conversation difficult if not impossible. Nonconscious processing, because of its short latency, finds the right words for us almost immediately and puts them in the right order according to the rules of grammar. It also decides for us which word or words should be emphasized in a conversation. All of this processing associated with making a conversation completes in a fraction of a second. Because of its quick processing time, the nonconscious mind makes conversation possible.[1]

Reading is another example. While reading a text we inadvertently use the nonconscious mind by not reading each letter but capturing an image of the whole word or a phrase at a time, this makes reading a lot faster. But the speed comes at the cost of accuracy, which is why we tend to ignore and even miss typos in the text. If we were to use the conscious mind for reading, it would take several minutes to read and understand just one sentence because we would have to consciously identify each letter, make a word, and then assemble words into a sentence and understand its meaning using the rules of grammar. But in that case, we would probably not miss typos! Incidentally it actually happens to some dyslexic individuals called letter-by-letter readers. These individuals have to read each letter before they can make a word and then a sentence. Obviously, their reading speed is slow—very slow. Letter-by-letter readers do not have visual or perceptual problems[2]; they are just not able to use the nonconscious mind that allows us to read by words and phrases rather than by letters.

Because nonconscious processing is quicker, the brain uses it to reduce the processing time of almost all conscious processes. That is why most conscious processes include components of the nonconscious mind. As discussed in Chapter 2, Nonconscious memory, in a series of experiments we demonstrated that during conscious recall information is first retrieved nonconsciously. It then goes through a second level of processing in the V3A-frontal loop. Only after this processing do we become consciously aware of the retrieval.[1,3,4] Conscious recall therefore involves two steps. In the first step, information is retrieved nonconsciously, and in the second step, it is processed to make the retrieval available to the conscious mind. Since conscious processes involve an additional step after nonconscious processing has finished, they are always slower than the conscious processes. In terms of speed the nonconscious mind is therefore, definitely smarter than the conscious mind.

Nonconscious processes are also smarter than conscious ones when all pieces of information needed to make a decision are not consciously available. Sometimes the nonconscious mind retains missing pieces of information and uses them to make a decision. That is the reason we tend to follow our "gut" when we cannot make a conscious decision. Gut feelings are nonconscious decisions. These decisions are quick but not always the best. As mentioned above, the primary objective of the nonconscious mind is to make quick decisions based on available facts and past experiences. Its quality depends on the quality of past experience. If we have limited experience with the situation, our nonconscious decision or the gut feeling is likely to be less than perfect.

Since nonconscious processing trades speed for accuracy, it is useful in situations that require quick decisions and actions. However, when accuracy is important conscious decisions are superior if all facts needed for making the decision are available. As discussed in Chapter 6, Decision making, when a decision involves working with a small number of variables, conscious decisions are better but if a large number of variables need to be processed, nonconscious decisions are probably better. Conscious processes have limited processing capacity because they require the attentional system, which cannot consciously process a large number of variables simultaneously.[5]

What happens if we do not know what variables or inputs are needed to make a decision or perform an action? For example, how do we know what and how much information is needed to decide whether to trust a person by looking at his or her face? Almost on a daily basis we encounter situations like this. Very often we are unsure about how much information we need to make a decision. In those situations the decision is made nonconsciously using our past experience with different people. The brain probably makes a profile of the "looks" of trustworthy and nontrustworthy people. Based on that profile the nonconscious mind makes a decision for us. That is why we remain unaware of the reason why we find some strangers trustworthy and others untrustworthy.

Normally, both the nonconscious and conscious mind are involved in the decision-making process. We tend to rely more on conscious decisions because we "know" or believe we know the pros and cons of different options and the reasoning behind those decisions. Since reasoning is not consciously known we are not confident about nonconscious decisions. Nonconscious decisions, however, boost our level of confidence in conscious decisions. If the conscious decision is different from the nonconscious one, we do not have a high level of confidence. This was shown by Bechara and colleagues in individuals with prefrontal damage.[6] These people were unable to make nonconscious decisions but could make conscious decisions about the nature of a deck of cards in the Iowa gambling task. Despite this, they picked cards from bad decks instead of good, because of the lack of confidence in their decisions. Nonconscious mind therefore boosts our confidence and motivates us to implement a decision.

The nonconscious mind also helps us when we do not have enough information to make a decision. That is what happens when making a decision to vote for a candidate in an election. Nobody knows whether a candidate will keep the promises he or she is making. Therefore it is difficult to make a conscious decision. Many investigators have studied this decision-making process and found that people make a choice based on their emotional connection with a candidate. This decision is primarily nonconscious. Once the nonconscious decision is made, the conscious mind creates a "justifiable" reason to boost the level of confidence and also to convince ourselves that the decision is based on issues rather than nonconscious preference.[7] This is why voters often change their minds several times before election, and after the election many have "buyer's remorse."

So, what is a good strategy? Accept the nonconscious decision (gut feeling) or go by the conscious "logical" decision? Obviously, there is no clear answer because the answer depends on a number of factors including a person's ability to make logical analysis, the variety and type of life experience, and the number of inputs involved in the decision-making process. As a rule of thumb, if you are sure that you have all the relevant facts to make a decision, and you know you can process those facts, then rely on conscious decision. As discussed above, reliability of nonconscious decisions depends on a number of factors. Most important being the nature and quality of life experience. Since nonconscious processes use past experience to predict the future, if a person has limited life experience, nonconscious processes (as well as conscious processes) may not have enough information to make a good decision. However, if information is not available to the conscious mind, nonconscious decisions may be better because the nonconscious mind may have the missing information.

Does that means the nonconscious mind is smarter? It depends on the situation but in general, because of faster processing, it is definitely smarter when a decision needs to be made quickly. Thus for making decisions in life-threatening situations the nonconscious mind is definitely smarter. It is also smarter when dealing with a large number of variables and in situations where making a quick decision is critical to performing a task. These tasks include conversation, reading, walking, and typing. When we have inadequate input for making a decision or when we are unsure about the nature of information needed to make a decision, the nonconscious mind again comes to our rescue and makes a decision for us.

Laboratory experiments suggest that the nonconscious mind makes better decisions when emotion is an important input or when the decision is based on subjective criteria. In one such experiment reported by Timothy Wilson of the University of Virginia a teacher rewarded students with a poster for work well done.[8] Half of them were given one of five posters while the other half were allowed to choose one from the same pool of five posters. A few weeks later the teacher asked if they still liked their posters. Most students who were not given a choice liked them but that was not the case with those who were given a choice. It is counterintuitive but most of them were not happy. A possible explanation could be their level of expectation. We set higher expectations for consciously made decisions as opposed to those made by someone else. Another reason for these results could be the lack of emotion. The conscious mind is not good at understanding emotional value in decision-making process. That is why consciously selected posters had no emotional content. In contrast, posters handed over to students by the teacher represented an emotional connection between the student and teacher/class. That is the reason why a trivial gift from a loved one feels more precious.

Since in certain situations nonconsicous deicisions are better than conscious ones, one should know whether a conscious decision is in agreement with nonconsiocus decision. How do we know if the nonconsiocus mind likes the decision? Most of the time, the nonconsiocus mind conveys its decision by making us comfortable or uncomfortable with a decision. As mentioned above, the nonconscious mind conveys its endorsement of or disagreement to a conscious decision by boosting or reducing our level of confidence. A real example of this situation is described in the book *Gut Feelings*[9] by German psychologist Gerd Gigerenzer (Fig. 12.1). A man nicknamed Harry loved two girls and was unable to decide who he should marry. He then used a logical approach and listed pros and cons of each girl and rated each with a positive and negative factor. Based on this approach he came up with a clear choice in favor of girl "A." However, he was uncomfortable with this choice, and so he married the other girl and apparently lived happily ever after. Obviously, the nonconscious decision in this case overruled conscious decision that was based on logic, but had no emotion. Since nonconscious processes are good at factoring emotion in the

FIGURE 12.1 Gerd Gigerenzer (1947) suggested that the rationality of decision making is based on many factors including cognitive limitations. *Reprinted from Wikimedia Commons.*

decision-making process, they helped Harry make the right decision. Paucity of emotional input is a limitation of the conscious decision-making process. This is why at times it falls short of our expectations.

Even though emotion is an important part of the decision-making process, it is unclear whether it always helps us make the right decisions. Traditional wisdom tells us not to make emotional decisions but we know emotions often help us make better decisions. Its effect on a decision probably varies depending on the type of decision. Emotions are important in decisions that are personal in nature like choosing friends, spouse, clothes, etc. Emotions provide a sense of security against uncertainty, which is why they are also important in situations where the impact of a decision is not immediately known. These decisions include career choice, voting for a candidate, buying a car, etc. It may take several months, even years before we know the impact of these decisions, which is why emotion plays an important role in decisions that affect our lives for a relatively longer period of time. In addition to offering security and comfort emotions also help us "connect with" and "own" the decision. Incidentally, emotion also decides how much effort we put in to implementing a decisions. Greater emotional involvement prompts us to put in more effort, which leads to a better outcome.

But decisions based solely on emotion may not be good. There are innumerable examples of emotional decisions going wrong in parenting and in the stock market, for example. Too much emotion limits efficiency and undermines objectivity. That is the reason surgeons do not perform surgeries on their family members and close friends. Emotional input needs to be used in moderation and judiciously to get maximum advantage without making it a liability. Thus while making a conscious decision we have to decide whether emotional input will make a decision better or worse. If it makes it better, then we have to decide how much weight emotion should have over the logic and other factors. Striking the right balance consciously is not easy. The nonconscious mind probably balances emotion and objectivity better (Fig. 12.2).

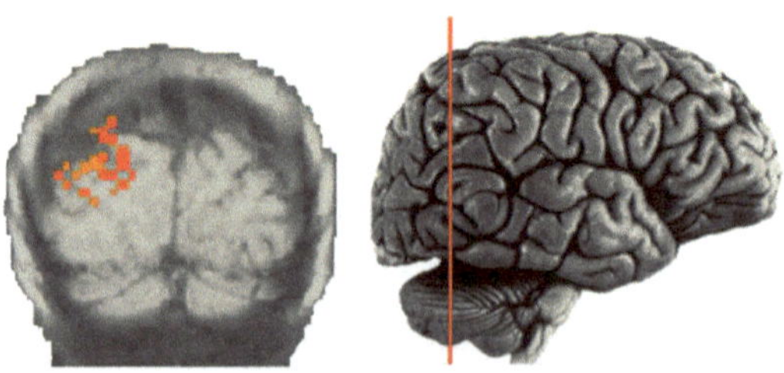

FIGURE 12.2 Area V3A involved in retrieval of nonconscious information. *Modified from Badgaiyan RD. Neuroanatomical organization of perceptual memory: an fMRI study of picture priming. Hum Brain Mapp 2000;**10**(4):197–203.*

Because conscious and nonconscious processes use different sets of information and different decision-making strategies, decisions would be better if we could selectively use components of both processes at will.[10,11] This could happen if we had a better understanding of the neural basis of decision-making process. Currently, we know almost nothing about the nonconscious decision-making process because it lets us know the decision but does not tell us what information or logic it used to make that decision. While the neural network involved in the decision-making process is not known, a small step in this direction was made in experiments studying the brain areas involved in information processing such as those discussed in Chapter 2, Nonconscious memory. In these experiments we established that area V3A, located at the junction of occipital, parietal, and temporal lobes (OPT junction) is critically involved in recall of nonconscious information. We proposed that this area has a gating mechanism to regulate release of information.[1,3,4,12,13] It allows release of a limited amount of information needed to accomplish a task. By limiting the amount of information it speeds up processing by limiting options and processing elements. It also avoids confusion by not flooding the brain with less relevant information. The released information remains nonconscious, until processed in the V3A-frontal loop, which allows back-and-forth communication between the two areas it connects (Fig. 12.3).[13,14] The information remains in this loop for about 400 ms before we become consciously aware of the retrieval. We do not know the mechanism by which processing in this loop brings information to conscious awareness, but it probably involves activation of the attentional and executive systems.[4] These systems process retrieved information to further limit options and eventually to select the best option for making a decision. As discussed in Chapter 6, Decision making, the anterior cingulate and prefrontal lobe are involved in this process.

The processing mentioned above involves a number of brain areas, and explains why we need time to make conscious decisions. Nonconscious decisions, on the other hand, are quick and sometimes instantaneous and do not involve these processing steps.

Nonconscious processing is similar to reflex action. Both are involuntary, fast, and their responses to a stimulus predictable. But reflexes are simple stimulus-response actions that do not require perception and higher-order processing. Nonconscious processing is much more complex. It involves cognitive processing at the highest level so that the brain understands significance of the stimuli and/or the situation. Further, it involves comparison of the current situation with similar situations encountered in the past, and track responses and outcome of those responses. Thus the nonconscious decision of "flight" from a potential life-threatening situation is based not only on understanding the threat but also on the response and outcome of previous encounters with a similar threat. It may also involve an evaluation whether the threat can be handled better by fighting or by running away. Normally

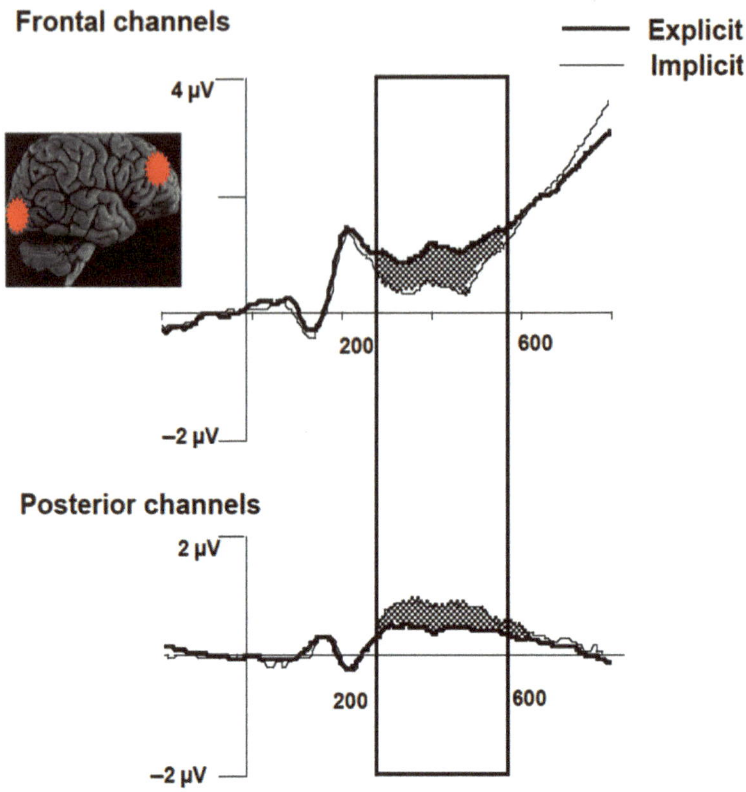

FIGURE 12.3 Contrast of the event-related potentials (ERPs) recorded during retrieval of studied words either consciously (explicit) or nonconsciously (implicit). It shows almost perfect overlap of activity in the frontal cortex and posterior cortical areas (including area V3A) during conscious retrieval. The overlap suggests two-way communication between the two areas to make nonconsciously retrieved information available to conscious awareness. Image of the brain shows approximate location of the frontal and posterior cortical areas from where potentials were recorded. *Modified from Badgaiyan RD. Conscious awareness of retrieval: an exploration of the cortical connectivity. Int J Psychophysiol 2005;**55**(2):257–62.*

several thousands, if not tens of thousands of neurons, are involved in this process. Since a stimulus takes about 0.5 ms to pass through a synapse (synaptic delay), processing through thousands of neurons would normally take several minutes, but nonconscious actions do not take that long. They are immediate. It suggests that nonconscious processing probably works as a collection of "reflex-like" actions. But how can a complicated task be carried out by simple reflex action? It is theoretically possible if the task is broken down into smaller events and there is a predetermined response for each of those events.

It has been suggested that the response to a stimulus is encoded in the basal ganglia as a "template."[15] The template is based on past experience and allows an immediate response, bypassing the usual steps needed for making a decision. I think these templates consist of modules that include information not only on the stimulus or event encountered but also past responses and outcome of those responses. When a similar event is encountered, the brain knows past experiences with the stimulus and immediately executes the most appropriate response. This strategy avoids reanalysis of the situation and weighing pros and cons of different options. If in the next encounter the event is perceived differently and requires a different response, the basal ganglia makes appropriate changes to the module and retrains the brain to initiate a different set of response. There are likely similar modules for motor actions that we perform repeatedly such as typing, walking, and talking.

Decisions made using these modules may not be the best because no two situations are exactly same. Therefore the brain might be using a module that closely resembles the current situation. Since we do not know how the brain defines "resemblance," there is an element of uncertainty and possibility of error. Since the conscious mind does not bypass processing steps, it takes more time but probably makes a better decision.

The nonconscious mind is smarter in situations that require quick decision and action, but when variables are well defined, are relatively few, and quick action is not needed, the conscious mind makes better decision. There is thus a place for both. They usually work together to help us make quick and good decisions. However, because of our inability to control the nonconscious mind, our level of confidence in nonconscious decisions is generally not very high. We could use nonconscious information at will if we had a better understanding of the neural network that controls the nonconscious decision-making process. We would then be able to manipulate the brain networks and combine the speed of the nonconscious and quality of the conscious decision-making processes to reduce the time need to make the best decisions and enhance our cognitive ability.

Bibliography

1. Badgaiyan RD, Posner MI. Time course of cortical activations in implicit and explicit recall. *J Neurosci* 1997;**17**(12):4904–13.
2. Warrington EK, Shallice T. Word-form dyslexia. *Brain* 1980;**103**(1):99–112.
3. Badgaiyan RD. Neuroanatomical organization of perceptual memory: an fMRI study of picture priming. *Hum Brain Mapp* 2000;**10**(4):197–203.
4. Badgaiyan RD. Conscious awareness of retrieval: an exploration of the cortical connectivity. *Int J Psychophysiol* 2005;**55**(2):257–62.
5. Dijksterhuis A, Bos MW, Nordgren LF, van Baaren RB. On making the right choice: the deliberation-without-attention effect. *Science* 2006;**311**(5763):1005–7.
6. Bechara A, Tranel D, Damasio H, Damasio AR. Failure to respond autonomically to anticipated future outcomes following damage to prefrontal cortex. *Cereb Cortex* 1996;**6**(2):215–25.

7. Westen D. *The political brain: the role of emotion in deciding the fate of the nation.* New York: PublicAffairs; 2007.
8. Wilson TD, Lisle D, Schooler JW, Hodges SD, Klann KJ, LaFleur SJ. Introspecting about reasons can reduce post-choice satisfaction. *Pers Soc Psychol Bull* 1993;**19**:331–9.
9. Gigerenzer G. *Gut feelings: The intelligence of the unconscious.* New York: Viking; 2007. p. 4–6.
10. Badgaiyan RD. Nonconscious processing and a novel target for schizophrenia research. *Open J Psychiatry* 2012;**2**(4A):335–9.
11. Badgaiyan RD. Nonconscious perception, conscious awareness and attention. *Conscious Cogn* 2012;**21**(1):584–6.
12. Badgaiyan RD. Cortical activation elicited by unrecognized stimuli. *Behav Brain Funct* 2006;**2**(17):1–5.
13. Schacter DL, Badgaiyan RD. Neuroimaging of priming: new perspectives on implicit and explicit memory. *Curr Direct Psychol Sci* 2001;**10**:1–4.
14. Badgaiyan RD, Schacter DL, Alpert NM. Auditory priming within and across modalities: evidence from positron emission tomography. *J Cogn Neurosci* 1999;**11**(4):337–48.
15. Wise SP, Murray EA, Gerfen CR. The frontal cortex-basal ganglia system in primates. *Crit Rev Neurobiol* 1996;**10**(3–4):317–56.

Chapter 13

Future outlook

It is clear in the previous chapters that the nonconscious mind is not a mere philosophical entity. Modern science is revealing its real neurobiological existence. It may surprise many that the concepts developed using objective scientific data do support some of the philosophical assumptions about nature and the scope of the nonconscious mind described over 4000 years ago. These concepts are also in agreement to an extent, with the modern philosophical theories, possibly because most of these theories are based on ancient texts.

As discussed in Chapter 1, Historical perspective, modern western philosophical theories of the nonconscious mind originated in a book published by German philosopher Karl Robert Eduard von Hartmann in 1869 (Fig. 13.1). The book was translated in English in 1884 as *Philosophy of the Unconscious: Speculative Results According to the Induction Method of the Physical Sciences.*[1] The concepts of the nonconscious mind proposed in this book were heavily influenced by Vedic philosophy. It describes three forms of the unconscious (nonconscious) mind: absolute unconscious, which is a substance of the universe and also the source of all other unconscious; physiological unconscious, which is at work at the origin, development, and evolution of living beings; and psychological unconscious, which lies at the source of our conscious mental life. This description is similar to the concept of *brahman* and *atman* described in the Vedas. *Brahman* refers to universal consciousness that transcends individual identity. *Atman* is the individual consciousness of each person and it is a part of *brahman*. The *Brihadaranyaka Upanishad* sums up this relationship in the sentence: *Aham brahmosmi* (I am Brahman), meaning I am part of *brahman*. *Atman*, or individual consciousness, has many components as discussed in Chapter 1, Historical perspective. It includes *sanskar*, which is the nonconscious mind that embodies the sum of all our life experiences. *Brahman*, *atman*, and *sanskar* are roughly the three forms of the unconscious mind described by Hartmann. These concepts with little variation comprise the basic framework of most modern philosophical theories.

Since philosophical study of the nonconscious mind has a much longer history, those theories are more advanced than scientific theories. Scientific studies have just begun and so far limited only to the study of individual

Neuroscience of the Nonconscious Mind. DOI: https://doi.org/10.1016/B978-0-12-816115-9.00013-9

FIGURE 13.1 Karl Robert Eduard von Hartmann (1842–1906). German philosopher who proposed the first modern western philosophical theory of the nonconscious mind. *Image from Wikimedia Commons.*

consciousness. Understanding the universal consciousness is a distant dream because we know nothing about it scientifically. Science has no evidence to prove its existence. However, absence of evidence is not the evidence of absence. It only means that currently available scientific tools and methods cannot prove its existence and scientific concepts are not advanced enough to understand the phenomenon.

Serious scientific study of consciousness began only a few decades ago. With advancement of science we may in the future be able to understand universal consciousness. Until recently the concept of nonconscious mind did not exist in the scientific domain because reliable evidence of its existence was not available. As discussed in previous chapters, now we have scientific evidence to prove its existence. We have developed experimental paradigms to elicit nonconscious functions and have identified some of the neural networks that control them. Study of nonconscious mind is no longer a taboo in science.

It is therefore, important to keep an open mind and not shut the door on a phenomenon that cannot be scientifically explained at this time. For this reason, hypnosis and ESP are included in this book. At this point we lack hard scientific evidence to explain either of these phenomena but we also know that our scientific journey of understanding the brain has just begun. Nonconscious processes were considered philosophical fantasies until just a few years ago—they were not considered scientific entities. Now we are trying to use the power of nonconscious mind to improve our physical and mental abilities. We can only guess what science may discover in the next few decades. The pace of scientific discovery will depend on our ability to study processes for which we have no adequate scientific explanation. If we close the door to these phenomena and discourage researchers to take on the challenge of understanding them, progress will be much slower.

To understand consciousness we will need significant advancement in our understanding of the neural networks that process information. At this time we have only basic knowledge of neuronal communication and know almost nothing about the way the brain communicates with the external world. This knowledge has not expanded beyond what mankind has known for centuries, that the brain receives signals from the external world through five senses: vision, hearing, smell, taste, and touch. With this level of knowledge, it is impossible to understand consciousness.

Our current knowledge about information processing at the neuronal level is primitive at best. While we do know how neurons communicate with each other, our knowledge has not advanced much in almost a century since discovery of chemical communication by Otto Loewi in 1921 (Fig. 13.4).

As discussed in Chapter 10, Extrasensory perception, there is evidence to suggest that the brains of two people communicate. But we have no clue how this happens. Similarly, we do not know how the brain communicates with matter, energy, and time. These are just a few examples of what we need to understand in order to build a scientific concept of both universal and individual consciousness.

At this point we actually do not even know what we do not know about the brain and consciousness. A few decades ago we were unaware of the existence of antibiotics and radio waves. Similarly, today we do not know what else exists beyond our sphere of knowledge. Thus, we need to keep studying the neural basis of known as well as unknown phenomena.

As mentioned above, understanding of universal consciousness, if it exists, would require us to know how the brain communicates with other brains, energy, matter, and time. Obviously, we have a long way to go before we can even validate the existence of universal consciousness or prove that it does not exist. If it does indeed exist and we are able to manipulate it, we could get the key to control our thought, cognition, emotion, behavior, destiny and probably our very existence. If we believe the wisdom of ancient thinkers, universal consciousness controls not only our activities but also those of other species. In short, universal consciousness is closest to the concept of God. Understanding of universal consciousness would be a huge discovery for mankind. It will, however, take decades if not centuries before we will acquire the ability to scientifically examine even the concept of universal consciousness. It would require development of not only methodologies to explore undiscovered aspects of brain function, but also a theoretical framework to understand the biological phenomenon.

Current technology allows us to detect only a few parameters of brain activity. These parameters include electrical potentials, magnetic field, hemodynamic response, chemical concentration, and metabolic activity. There is no reason to believe that these parameters capture all brain activities. In the last decade, it was discovered that the neurons respond to visual and auditory signals,[2–5] but the significance of these signals is unclear. There are probably many other modes of neuronal communication we do not know about.

Additionally, at this time we are able to detect only gross changes in some of the parameters that measure brain activity. Changes that are weak or confined to a small group of neurons cannot be detected in live human brain using noninvasive techniques. Therefore, we have a long way to go before we are able to understand brain processing. But that is not surprising because modern neuroscience has a relatively short history. Until about 100 years ago we had no clue about how the brain works. When people began to think about it, there was excitement as well as confusion. The confusion led the Nobel committee in 1906 to award Nobel prize to people having diametrically opposing views about structure and function of the brain. That year, it was awarded jointly to Camilo Golgi (Fig. 13.2) and Ramon y Cajal

FIGURE 13.2 Camilo Golgi (1843–1926). Proponent of the reticular theory. *Reproduced from Wikimedia Commons.*

FIGURE 13.3 Ramon y Cajal (1852–1934). Proposed the neuron doctrine. *Image reproduced from Wikimedia Commons.*

(Fig. 13.3). In his Nobel lecture Golgi supported the reticular model of nervous system. This model, originally developed by German anatomist Joseph von Gerlach, suggests that the brain works as a single continuous network. Cajal refuted this theory and talked about his neuron doctrine. He rejected reticular theory and suggested that the brain consists of distinct units of cells called neurons that communicate with each other to accomplish different tasks.

Even though Cajal conceptualized neuronal communication in the early 1900s, it was not known for sure how neurons communicate. There was speculation about bioelectrical or chemical mode of communication but no evidence until a German pharmacologist Otto Loewi (Fig. 13.4) demonstrated in 1921 that neurons control heart rate using chemical signals. In a famous experiment, he dissected two frogs and exposed their hearts. He then electrically stimulated the vagus nerve of one animal and noted slowing of the heart. He then collected fluid released during stimulation and put it on the heart of a second frog. The fluid caused heart of the second frog to slow down suggesting that the chemical released in the first animal slowed the heart of the second frog. This experiment convinced people that neurons communicate using chemicals.

The next breakthrough came almost half a century later in 1959 when two researchers of University College of London, E.J. Furshpan and D.D. Potter, discovered a second mode of communication. They demonstrated in crayfish that neurons also communicate using electrical signals.[6] In the last

FIGURE 13.4 Otto Loewi (1873–1961). Demonstrated chemical transmission of neural signals. *Reproduced from Wikimedia Commons.*

70 years, no other means of neuronal communication have been discovered. While recent evidence suggests that neurons respond to photic and auditory signals,[2–5,7,8] it is not clear whether these signals are used for neuronal communication. Hundreds of chemicals are now known to influence brain activity and new chemicals are still being discovered. Thus, our knowledge of communication channels of the brain is incomplete.

Some of the media the brain maybe using to communicate with its different areas and with outside world could be beyond our ability to detect. We know that our sensory systems can detect only a narrow band of electromagnetic and compression waves. Our visual system detects as light electromagnetic waves with wavelengths between 400 and 720 nm. The auditory system can detect compression waves with wavelengths between 17 m and 17 mm (frequency between 20 Hz and 20 kHz). In nature there are electromagnetic and compression waves that are way beyond the range our sensory systems are capable of detecting. Some of these waves can now be detected using radio and television (Fig. 13.5), but what about frequencies beyond the range we can detect either by our sensory system or by the devices we have built so far? Devices that can detect the electromagnetic waves our senses cannot detect were developed only in the last century. Therefore, much work is still to be done.

Since we can detect signals only within a limited bandwidth, our senses are "blind" to a lot of information out there. Some animals can detect signals

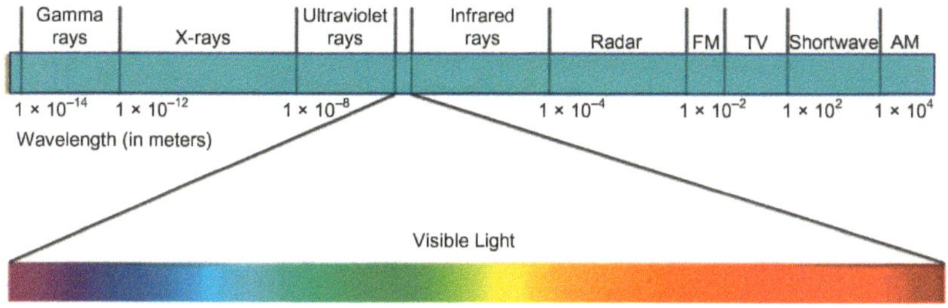

FIGURE 13.5 Electromagnetic waves we can detect either by our visual system (Visible Light) or by special devices.

outside the range humans can detect. Those animals are thus better informed about their surroundings than we are. With the ability to detect ultraviolet light (wavelength 10–400 nm) insects and butterflies may have a different view of the world than we do. Similarly, snakes, with their ability to detect infrared light (wavelength 700 nm–1 mm), and whales, elephants, and dogs with their ability to detect compression waves far beyond our auditory range, may also have different perspectives of the world. This explains why the behavior of some animals changes before the onset of natural disasters like earthquakes, tsunamis, and volcano eruptions. They can obviously sense something that we cannot. One way to address this human handicap is to develop devices that can detect these undetectable signals. Detection of those signals could help us understand not only some of the unexplained phenomena, but also our own mind and the world around us.

Since we can detect only a small fraction of the signals we encounter every day, how do our bodies react to signals that cannot be detected? Some of those signals like gamma rays and X-rays damage the body. For other signals, there are several possibilities. Our bodies could ignore them or detect them in a way that our brain cannot analyze or interpret. Is it possible that some of those signals result in "gut feelings," or could let us interact with energy, time, other people, and animals? We don't know.

Is it possible that only some of us can detect "undetectable" signals and make sense of them? We know some of us can see light that is outside our normal range of vision and there are people who can hear sounds that others cannot. Similarly, people who have had their cornea removed and replaced by artificial prostheses can see ultraviolet light and under certain conditions some people can see normally invisible infrared light.[9] Thus, it is very possible some of us can detect signals that others cannot. Could those signals be responsible for the telepathic ability of some individuals? There is no reason to believe why it is not possible.

We know that our bodies sense a variety of internal signals and react to them even though we are consciously unaware of their responses. For example, we do not know when baroreceptors located inside our blood vessels are activated. These receptors constantly monitor blood pressure and help keep it within a normal range. Similarly, there are chemoreceptors that detect levels of different chemicals in the blood and other body fluids. Peripheral chemoreceptors called aortic and carotid bodies detect carbon dioxide concentration and based on the concentration corrective action of hyper- or hypoventilation is initiated automatically without our conscious knowledge. A number of other chemoreceptors detect levels of hormones and chemicals. Proprioceptors sense the positions of joints. The activity of these receptors is controlled by the nonconscious mind, which probably also monitors corrective responses initiated by them.

Dozens of receptors that detect different signals have been isolated in the human body and new receptors are still being discovered. Each of these receptors detects a specific signal, which can either be a chemical or physical entity. Signals that activate many, but not all of those receptors are known. Signals that activate a large number of these receptors are not known. These receptors are called silent receptors. We do not know what activates silent receptors, but this does not mean they do not detect anything. Is it possible that some of these receptors detect metaphysical events? Obviously, this possibility cannot be discounted. Additionally, there are many receptors that have not yet been discovered. It is difficult to predict what those receptors might be detecting and what abilities do they impart on us when activated.

Logic would suggest that if known receptors account for all of our abilities, there are no more receptors to look for. But what about silent receptors? These receptors are there for a reason. So do we have abilities we do not know about? Quite possibly yes. We know some of us possess extraordinary abilities, and there is a wide variation in our ability to detect signals. But we do not know why some of us are better than others at detecting and analyzing signals. Musicians, for example, are better than most of us at detecting changes in musical notes. Do they get this ability by enhancing the efficiency of existing receptors or by using some of the silent receptors we do not normally use? We do not know for sure. There is another possibility. May be all of us have these abilities but only a few can bring them to conscious awareness. As discussed earlier, we have no conscious awareness of many functions the brain performs. In order for us to become consciously aware of an information it needs to be processed in the V3A frontal loop as discussed in Chapter 2, Nonconscious memory. It is possible that some of us are able to channel this information through the V3A frontal loop and become consciously aware of the "extraordinary" abilities we all possess but are unable to bring to conscious awareness.

Some of these abilities could include the abilities that make many less evolved animals smarter than us. For example, the geographical localization ability that many migratory animals possess. Some of these animals travel up to 20,000 kilometers and do not get lost. They return to the exact location of origin. We do not know for sure what guides them and how do they precisely locate geographical landmarks. It is believed that these animals are guided by the earth's magnetic field. But it is not clear whether humans lack this ability or if it is dormant. Disuse atrophy of organs, organ systems, and abilities is a well-known evolutionary process. It is also possible that migratory animals communicate with the matter to locate their position. For this communication, they maybe using a medium or media that we do not know about.

It is thus possible that some of our abilities are dormant and unknown to us. Functions like ESP could be one of those hidden abilities. Since these abilities remain at the nonconscious level, we are not able to use them at will. We do not even know if we have those abilities because information retained in or acquired by the nonconscious mind does not get into our conscious awareness, unless it is processed in the V3A frontal loop. Not all information retained in the brain goes through this additional processing, which is why we do not know what is retained in our nonconscious mind. So, if we do possess ESP at the nonconscious level we cannot use it at will until we acquire the ability to control nonconscious mind. We are not there yet.

Modern neuroscience has a relatively short history of about 100 years. It is only at the threshold of understanding human brain and its abilities. We have absolutely no knowledge of many brain processes and functions. It is thus naïve to rule out existence of functions that have not yet been discovered because there are many functions waiting to be discovered. The pace of discovery can be enhanced by keeping an open mind and proactively investigating phenomena and abilities that are still unexplained. These studies will also prompt us to develop newer technologies and tools to study the brain.

Since the brain uses multiple channels for internal communication and also to communicate with the outside world, it is important to be able to understand those channels. We have begun to discover some of them but the existence of silent receptors suggests that many signals we are capable of detecting have not yet been discovered. We donot know whether those signals are used by the brain for internal or external communication.

Considering that we had no tool to study the live human brain until about 100 years ago, it is not surprising that we do not have the ability to detect all brain processes. The very first tool to study the live human brain was developed in 1924 by a German physiologist Hans Berger (Fig. 13.6) who recorded the electrical activity of brain using EEG (electroencephalography), which is now used extensively to diagnose brain disorders. Discovery of EEG was initially received with skepticism because Berger was unable to explain the physics of brain activity he was recording. It was not recognized as a genuine brain signal until 1937 when scientists agreed that signals Berger was recording indeed came from the brain.

After its discovery and recognition as a valid brain signal, the EEG remained the only tool to study live human brain for decades. Then, an instrument to detect changes in the magnetic field of the brain was developed by David Cohen in 1968 at the University of Illinois. This technique is called

FIGURE 13.6 Hans Berger (1873–1941). Recorded electrical activity of the brain using EEG for the first time in 1924. *Image modified from Wikimedia Commons.*

magnetoelectroencephalography (MEG). Shortly thereafter, positron emission tomography was developed in 1975 to detect changes in metabolic activity of the brain. In the 1980s EEG was modified to allow recording of task-related changes in the brain called event-related potential (ERP). This technique is extensively used to study the temporal sequence of changes in electrical potential during cognitive and behavioral processing. The next advancement happened about a decade later in 1990 when functional magnetic resonance imaging (fMRI) technique was developed at Bell Laboratories by Seiji Ogawa.[10] This technique takes advantage of differences in the magnetic fields created by oxygenated and deoxygenated blood. By analyzing this difference, fMRI detects task induced changes in hemodynamic response, which indirectly reflects blood flow in that area. This is an indirect measure of the activity of a brain region. Currently this technique is extensively used to study the human brain and its functions. Around the time the fMRI technique was developed, a British physicist named Anthony T. Barker and colleagues developed transcranial magnetic stimulation (TMS) at Sheffield University.[11] TMS allows us to activate or deactivate specific cortical areas via electrodes placed over the scalp. At the turn of the last century scientists were able to detect task-induced changes in electrical potentials, magnetic fields, metabolic activity, and hemodynamic response using various neuroimaging methods. Surprisingly, at that time there was no tool to detect task-induced acute changes in concentration of neurochemicals.

Even though, as discussed earlier it was established beyond doubt that the nervous system uses chemical agents for internal communication. Therefore, in 2003 we developed a technique to detect and map neurotransmitters released acutely in the live human brain during performance of a task.[12–20] This technique, called neurotransmitter imaging or single-scan dynamic molecular imaging technique (SDMIT) extended the scope of neuroimaging research and allows the study of this important aspect of brain activity.

Despite rapid advances in technology in the last four decades, we are still unable to detect all parameters that measure brain activity. Moreover, each neuroimaging technique mentioned above detects only one aspect of the activity. Thus, EEG or ERP techniques can only detect changes in electrical potentials. Therefore, if a brain activity induces changes only in chemical milieu, these techniques will not be able to detect. Detection of only one aspect of brain activity provides limited information, which in turn limits our understanding of the brain processes. Limited information could even be deceptive because incomplete knowledge often leads to incorrect conclusion. Since we are trying to formulate rules based on the data, it is important that we do not depend on the data that reflects only one aspect of the brain activity. That is the reason, some investigators are working on multimodal imaging that uses multiple techniques simultaneously. This approach could provide more reliable data but progress in this direction is rather slow because of the inherent problems of using multiple imaging techniques simultaneously. For example, fMRI uses strong magnetic fields to detect changes in hemodynamic response and techniques like MEG and EEG require magnetically shielded environment. Therefore, it is a challenge to use these techniques simultaneously. While these problems will likely soon be resolved, multimodal imaging is only a small step. The real difficulty is finding the undiscovered signals and media of communication the brain uses. Once they are discovered then the next step is to develop tools to detect them.

Even if we develop tools to detect all activities of the brain, we may not be able to understand its functions and capabilities because we do not have theories to explain brain activities. We have to develop theories that explain biological processes. Those theories will help us understand significance of the data acquired using either classical or newer techniques.

Currently, we borrow theories from physics and chemistry to understand biological processes. Since these theories are developed to explain behavior of nonliving objects, they can explain only certain aspects of the biological processes. They cannot explain uniquely biological phenomena like consciousness, qualia, self, volition, creativity, etc. Lack of biological theories is an important reason we have poor understanding of these phenomena. We need these theories to understand processes that are unique to life. Even though theories of physics and chemistry gave us the ability to analyze physical and chemical properties of living beings, we have not been able to create a single living cell using those theories. We know the chemical composition of living organisms but life is not created by mixing those chemicals. We can create organs from stem cells but cannot create stem cells themselves. Life therefore is much more than physics and chemistry.

Borrowed theories do not help us understand complicated brain processes either. Using physical laws we know how stimulation of motor area of the brain causes muscles to move. We also know how signals are transmitted from the brain to muscles and how muscles contract to make movement possible. But we do not know how we activate motor areas just by having the will to move. And how do we know which part of the motor cortex we must activate to make a desired movement. Not only that, we know that the cerebellum, basal ganglia, somatosensory areas, and host of other brain areas are also involved in making movements smooth and precise, but how do we activate those areas without understanding the process? We perform complicated tasks very efficiently, without knowing how the brain controls those processes.

On top of the above borrowed theories, we also apply borrowed mathematical methods to validate biological data. These methods are also not designed to explain biological processes. For example, the mean is by far the most common measure used in biology. Since this measure was developed and validated for nonliving objects, it does not accurately measure biological processes. Thus, if a machine randomly puts six balls into six empty slots, we can reasonable expect to find the same number of balls in each slot after repeated trials. However, if we try to create a similar situation in a biological setting and imagine six people waiting for a train at midnight at a station with a train pulling in with six empty cars. It is highly unlikely that each empty car will have the same number of passengers, no matter how many times the situation is repeated. Most likely all six people will board the same car. Thus, there is a very small chance the statistical mean of one passenger per car will be true. But that is what we expect in biological processes. Since, currently used statistical methods do not help us develop biological data and theories, we need to develop methods that take into consideration the uniqueness of biological processes and the nature of living beings.

Another problem that has slowed scientific discovery of the brain and probably other sciences as well is the rigidity of our thought process. Ironically, "scientific thinking" instead of encouraging free thinking limits the range of our thinking by teaching to believe in theories and to think only in certain ways. It limits our imagination to science's rigid framework and discourages exploration of concepts outside a predefined parameter. For example, we are taught that one plus one is always two. But as we know, in nature, this is not always the case. There are many examples that defy basic "scientific concepts." Thinking within a strict scientific framework prevents the "out of the box" thinking that is necessary for rapid advancement of knowledge and for understanding unexplored areas. There is thus a need to break away from that rigidity and encourage young scientists to let their imaginations fly free. As discussed in Chapter 8, Creativity, the ability to make unusual associations is key to creating new ideas and for making groundbreaking discoveries. Unfortunately, "scientific thinking" discourages making unusual connections. If scientists are encouraged to think beyond the parameters set by science, in the near future we should be able to understand most of the brain's functions. Those functions might include conscious and nonconscious processes.

Once that happens, we will be able to control our own mind and enhance cognitive and physical abilities to the extent we cannot even imagine today. It will be a game changer for species *Homo sapiens*. Until that happens, we have to make a guess about depth of the ocean while dabbling at seashore.

Bibliography

1. Hartmann EV, Coupland WC. *Philosophy of the unconscious; speculative results according to the inductive method of physical science.* new ed. K. Paul, Trench, Harcourt, Brace & Company; 1931.
2. Borrell J, Torrellas A, Guaza C, Borrell S. Sound stimulation and its effects on the pituitary-adrenocortical function and brain catecholamines in rats. *Neuroendocrinology* 1980;**31**(1):53–9.
3. Chaudhury S, Nag TC, Jain S, Wadhwa S. Role of sound stimulation in reprogramming brain connectivity. *J Biosci* 2013;**38**(3):605–14.
4. Gysbrechts B, Wang L, Trong NN, Cabral H, Navratilova Z, Battaglia F, et al. Light distribution and thermal effects in the rat brain under optogenetic stimulation. *J Biophotonics* 2016;**9**(6):576–85.
5. Zhang J, Laiwalla F, Kim JA, Urabe H, Van Wagenen R, Song YK, et al. Integrated device for optical stimulation and spatiotemporal electrical recording of neural activity in light-sensitized brain tissue. *J Neural Eng* 2009;**6**(5):055007.
6. Furshpan EJ, Potter DD. Transmission at the giant motor synapses of the crayfish. *J Physiol* 1959;**145**(2):289–325.
7. Szobota S, Isacoff EY. Optical control of neuronal activity. *Ann Rev Biophys* 2010;**39**:329–48.
8. Ibsen S, Tong A, Schutt C, Esener S, Chalasani SH. Sonogenetics is a non-invasive approach to activating neurons in *Caenorhabditis elegans*. *Nat Commun* 2015;**6**:8264.
9. Palczewska G, Vinberg F, Stremplewski P, Bircher MP, Salom D, Komar K, et al. Human infrared vision is triggered by two-photon chromophore isomerization. *Proc Natl Acad Sci USA* 2014;**111**(50):E5445–5454.
10. Ogawa S, Lee TM, Kay AR, Tank DW. Brain magnetic resonance imaging with contrast dependent on blood oxygenation. *Proc Natl Acad Sci USA* 1990;**87**(24):9868–72.
11. Barker AT, Jalinous R, Freeston IL. Non-invasive magnetic stimulation of human motor cortex. *Lancet* 1985;**1**(8437):1106–7.
12. Badgaiyan RD, Sinha S, Sajjad M, Wack DS. Attenuated tonic and enhanced phasic release of dopamine in attention deficit hyperactivity disorder. *PLoS One* 2015;**10**(9): e0137326.
13. Badgaiyan RD. Imaging dopamine neurotransmission in live human brain. *Prog Brain Res* 2014;**211**:165–82.
14. Badgaiyan RD. Detection of dopamine neurotransmission in "real time". *Front Neurosci* 2013;**7**:125.

15. Badgaiyan RD, Wack D. Evidence of dopaminergic processing of executive inhibition. *PLoS One* 2011;**6**(12):e28075.
16. Badgaiyan RD. Neurotransmitter imaging: basic concepts and future perspectives. *Curr Med Imaging Rev* 2011;**7**:98–103.
17. Badgaiyan RD. Dopamine is released in the striatum during human emotional processing. *Neuroreport* 2010;**21**(18):1172–6.
18. Badgaiyan RD, Fischman AJ, Alpert NM. Dopamine release during human emotional processing. *Neuroimage* 2009;**47**(4):2041–5.
19. Fischman AJ, Badgaiyan RD. Functional imaging of neurotransmission. *Curr Med Imaging Rev* 2007;**3**:385–403.
20. Badgaiyan RD, Fischman AJ, Alpert NM. Striatal dopamine release during unrewarded motor task in human volunteers. *Neuroreport* 2003;**14**(11):1421–4.

Index

Note: Page numbers followed by "*f*" refer to figures.

E

J

K

L

M

N

FSC
www.fsc.org
MIX
Papier | Fördert
gute Waldnutzung
FSC® C083411

Kolibri 360 GmbH
Ferdinand-Jühlke-Straße 7
99095 Erfurt, Deutschland
produktsicherheit@kolibri360.de